Springer Theses

Recognizing Outstanding Ph.D. Research

Aims and Scope

The series "Springer Theses" brings together a selection of the very best Ph.D. theses from around the world and across the physical sciences. Nominated and endorsed by two recognized specialists, each published volume has been selected for its scientific excellence and the high impact of its contents for the pertinent field of research. For greater accessibility to non-specialists, the published versions include an extended introduction, as well as a foreword by the student's supervisor explaining the special relevance of the work for the field. As a whole, the series will provide a valuable resource both for newcomers to the research fields described, and for other scientists seeking detailed background information on special questions. Finally, it provides an accredited documentation of the valuable contributions made by today's younger generation of scientists.

Theses are accepted into the series by invited nomination only and must fulfill all of the following criteria

- They must be written in good English.
- The topic should fall within the confines of Chemistry, Physics, Earth Sciences, Engineering and related interdisciplinary fields such as Materials, Nanoscience, Chemical Engineering, Complex Systems and Biophysics.
- The work reported in the thesis must represent a significant scientific advance.
- If the thesis includes previously published material, permission to reproduce this must be gained from the respective copyright holder.
- They must have been examined and passed during the 12 months prior to nomination.
- Each thesis should include a foreword by the supervisor outlining the significance of its content.
- The theses should have a clearly defined structure including an introduction accessible to scientists not expert in that particular field.

More information about this series at http://www.springer.com/series/8790

Maxime J. Jacquet

Negative Frequency at the Horizon

Theoretical Study and Experimental Realisation of Analogue Gravity Physics in Dispersive Optical Media

Doctoral Thesis accepted by
the University of St. Andrews, St. Andrews, UK

 Springer

Author
Dr. Maxime J. Jacquet
Faculty of Physics
University of Vienna
Vienna
Austria

Supervisor
Prof. Friedrich König
University of St. Andrews
St. Andrews
UK

ISSN 2190-5053   ISSN 2190-5061   (electronic)
Springer Theses
ISBN 978-3-030-08170-6   ISBN 978-3-319-91071-0   (eBook)
https://doi.org/10.1007/978-3-319-91071-0

Printed on acid-free paper

This Springer imprint is published by the registered company Springer International Publishing AG part of Springer Nature
The registered company address is: Gewerbestrasse 11, 6330 Cham, Switzerland

Supervisor's Foreword

The two outstanding physical theories of the twentieth century are the theory of general relativity and the theory of quantum mechanics. Both of these have transformed our understanding of nature. General relativity is associated with large scales (the cosmos) and quantum mechanics is considered valid on small scales (the atom). They have led to remarkable success such as the understanding of binary pulsar dynamics or the detection of gravitational waves for relativity and the calculation of atomic orbits and the discovery of nonlocality for quantum mechanics. These theories are, however, more than merely approximations at different scales and as such do not offer an absolute length scale. In fact, they are notoriously difficult to combine without inevitable paradoxes. The question arises, whether quantum mechanics and general relativity could play a role on equal length scales simultaneously and which theory would describe this. Black holes are often portrayed as combining both sides, and thus as a benchmark for future theories of quantum gravity.

Black holes are a consequence of Einstein's theory of general relativity from 1915. Within months, Schwarzschild found that this theory would predict a gravitational field of a point mass, a black hole. The gravitational field in the vicinity of this point mass would be strong enough to capture all particles, even photons, i.e. light. The range over which the particles are captured is limited by the event horizon, which separates the—possibly—escaping fields from the fields falling into the black hole. Thus, the event horizon signifies the size of the black hole. Outside, the object would not emit light and be 'black'.

This was how black holes were viewed until 1974, when Stephen Hawking considered quantum fields around the event horizon. In a groundbreaking paper, Hawking showed that quantum effects would lead to the emission of particles from black holes, 'Hawking radiation'. This effect would be fuelled by the mass of the black hole and even intensify as the black hole mass reduces, leading to a final explosion of particles in the last phase of 'black hole evaporation'. At last, an object was found that leads to profound physical effects if the two theories are combined. Until now, Hawking's prediction has not been experimentally verified. The universe is a large laboratory, but we cannot isolate a black hole to perform measurements. In fact, it turns out that the cosmic microwave background is far stronger than Hawking radiation. Although the radiation exists and can be detected efficiently, the background radiation is too strong.

In physics, the observable effects are described by equations. Each equation, however, can describe different physical systems, and so it is possible to transfer the same physics from one system to another. In 1981, Bill Unruh discovered that fluid flow of water can simulate an event horizon and, in consequence, must emit 'analogue Hawking radiation' in the form of sound waves. Although this particular system is impractical too, it demonstrated in principle that Hawking radiation can be detected in a terrestrial laboratory without background noise. Since, various methods have been proposed for the production of analogue Hawking radiation in a variety of systems, such as Bose–Einstein condensates, polariton condensates or nonlinear optics. After all, the detection of the elusive Hawking radiation seems within reach.

A realisation in optics leads to the highest expected particle emission, the highest effective temperature and no thermal background. Maxime's thesis is devoted to this elegant variant of analogue Hawking radiation. The thesis is presented in a form accessible to interested undergraduate and postgraduate students. Maxime is a passionate science communicator and gives a full account of the topic in an introductory and clear way. Starting with the basis of special relativity, the thesis introduces the relevant concepts of general relativity, focusing on wave motion at the horizon. In particular, he introduces the thermodynamic physics of black holes and derives Hawking's seminal result. The thesis then systematically introduces a field-theoretical framework for optical analogues, including the kinematics governing the fields, for light–matter interaction in a dispersive medium. This model allows for an analytical calculation of the all-important conversion of vacuum fluctuations into Hawking radiation. For the first time, coupling to non-optical branches and all kinematic configurations are included in the description. Based on this, emission spectra and intensities are calculated which give unprecedented insight into the emission from a highly dispersive system, both from a theoretical as well as experimental point of view. Maxime also introduces the relevant concepts for fibre-optical analogues, such as soliton formation and propagation, as an experimental means to produce optical horizons.

In an experimental part, the thesis develops a clear and systematic way to experimentally approach the problem. Based on the formalism developed in the thesis, Maxime analyses the possibility to obtain an experimental signal of detectable strength, including aspects of spectral resolution and quantum efficiency. He then demonstrates the construction of an experimental setup and measurements of unprecedented sensitivity in the search for stimulation of the Hawking effect.

Major parts of the thesis were presented at key international conferences receiving great interest. The thesis itself was refereed by the 'founding father' of the field, Prof. William Unruh, Vancouver. Optical Hawking radiation has become more tangible due to Maxime's work and it is exciting times to possibly witness the first detection of optical Hawking radiation, unfortunately too late for Stephen Hawking. Yet, it is more than satisfying to see the enthusiasm of a new generation of future academics.

We gratefully acknowledge the support by the EPSRC for this research.

St. Andrews, UK   Dr. Friedrich König
March 2018

Abstract

Quantum vacuum fluctuations on time-dependant-curved spacetimes cause the emission of particles. This effect results from the mixing of positive and negative frequency waves, and its most famous instance is Hawking radiation from black holes. Unfortunately, the latter cannot be observed in astrophysics, because of its ultra-low temperature. This thesis considers the problem of recreating the physics of wave motion on curved spacetimes in the laboratory so as to enable the study the scattering of waves at the event horizon. Laboratory analogue gravity systems typically are inhomogeneous, dispersive media in which the velocity of waves may be controlled to mimic the effects of spacetime curvature. For example, this can be realised in optics. Here, I present an analytical description of spontaneous emission in optical analogues. I consider a moving refractive index perturbation in an optical medium, which exhibits optical event horizons. Based on the field theory in curved spacetime, I formulate an analytical method to calculate the scattering matrix that completely describes mode coupling leading to the emission of photon pairs in various kinematic configurations. I apply the method in a case study, in which I consider a moving refractive index step in bulk-fused silica. I calculate the emission spectrum, which is a key observable, in the moving frame as well as in the laboratory frame. I find that emission from horizons is characterized by an increased photon flux and a signature spectral shape. In particular, the spectra are dominated by a negative frequency mode, which is the partner in any Hawking-type emission. This is interesting as it has never been observed either theoretically or experimentally before. An experiment aimed at stimulating the emission into negative and positive frequency modes is assembled. The classical effect of mode conversion in this optical scheme is clearly shown to be a feature of horizon physics. These theoretical and experimental methods and findings pave the way to the observation of particles emitted from the event horizon by the Hawking effect in dispersive systems.

Acknowledgments

To endless conversations. And friendship.

Upon finishing this thesis, my first thoughts and thanks go to Friedrich König, who welcomed me by his side and spent the past 45 months sharing his research, thoughts, findings and insights with me. Any attempt to list or summarise all I got and benefited from his supervision would be vain, for such assessment will only be possible with hindsight. Thank you, Frieder, for making a Physicist out of me.

As a young researcher, I was always welcomed by any member of the event horizon community I turned to. In particular, I am intellectually indebted to Stefano Finazzi, Ralph Schützhold, Bill Unruh, Silke Weinfurtner, Germain Rousseaux, Daniele Faccio and Iacopo Carusotto who were open to conversing with me. Many aspects of the thesis I wrote stem from those interactions. Special thanks go to Goëry Genty who took me aboard whilst in Varenna. In writing his lecture notes, I learnt a lot of nonlinear fibre optics.

I also wish to thank those who shared this studentship with me, accompanied me for some parts of the journey and made those past years the most enjoyable time possible—in order of appearance: Chris, Philipp, Larissa, Laura, Isabelle, Maya, Francisco, Myria and Ivana.

J'ai toujours été encouragé à être curieux par mes mère, père et beaupère, Laetitia, Gilles et Frédéric, et inspiré par l'exemple de mon grand-père, Yannick, il n'y a pas de plus de belle idée qui puisse être inculquée. Merci.

My studentship was partly funded by a 600th Anniversary Fellowship, and EPSRC.

Cette thèse, l'écrit, le moment et l'aventure intellectuelle sont dédiés à Kahina.

Contents

Chapter 1
Introduction

Light waves in media can be made to propagate on an effectively curved spacetime. Such *optical spacetimes* are curved Lorentzian manifolds which enable the study of some features of gravity in the laboratory.

Importantly, in the experiment, the effective curvature of the spacetime is not created by gravity itself[1]: this is done by modifying the refractive index experienced by light upon propagation in the medium, I will explain how later in this chapter. As a result, the dynamics of optical spacetimes are not determined by the Einstein equations [1, 2] but are implicit in Maxwell's equations [3–5]. The existence of these curved spacetimes *without gravity* is phenomenal, for they enable theoretical and experimental investigations of uttermost fundamental importance. Their study requires us to think carefully about the distinction between kinematics and dynamics in General Relativity. In turn, this opens a new perspective on the connections and differences between Lorentzian geometry, Einstein's equivalence principle, and full General Relativity.

Optical spacetimes are one example of the rich family of laboratory-based *analogue gravity* systems, in which waves are made to propagate on effectively curved spacetimes [6]—sibling systems range from fluids to superconducting circuits. Thanks to these laboratory analogues, we may access generic features of gravity (that one would usually think of as intrinsically aspects of gravity) which are both classical and semiclassical in nature: typically (but not exhaustively), it is possible to study event horizons, and the spontaneous emission of quanta at those horizons by the Hawking effect [7]. In this Thesis, I use the optical scheme to investigate generic features of curved spacetimes and Quantum Field Theory in curved spacetimes.

In this introductory chapter, I will briefly sketch the history of the science of analogue gravity systems. I will review the main physical arguments that motivate and support this science, and highlight some of its most fascinating and surprising

---

[1] Note that the real spacetime structure in any Earth-based laboratory is approximately Minkowskian.

© Springer International Publishing AG, part of Springer Nature 2018
M. J. Jacquet, *Negative Frequency at the Horizon*, Springer Theses,
https://doi.org/10.1007/978-3-319-91071-0_1

peculiarities. This will arm the reader with the fundamental ideas of the analogue gravity research programme, and of the work done in this Thesis. Let us begin with the historical motivation behind the study of analogue gravity: understanding how vacuum fluctuations in the vicinity of black holes lead to the spontaneous emission of quanta by the Hawking effect.

The Hawking effect as the mixing of positive and negative frequency waves

Black holes are believed to result from the gravitational collapse of stars in on themselves [8]. These bodies are so compact that the escape velocity from their surface (this boundary is called the *event horizon* [9]) is greater than the speed of light, hence their name. Nothing, even light, can come out of a black hole. Although direct observation of black holes is possible via their influence on the curvature of spacetime—they create gravitational waves when they orbit around a point, and emit energy in the form of gravitational waves upon merging, as was detected by the LIGO-Virgo Collaboration in 2016 [10]—the region of spacetime behind the event horizon remains hidden from us. The event horizon appears to be a one-way door: beyond it, motion can only be directed toward the centre of the black hole, where a singularity lies [8]. This implies that black holes cannot emit heat. And yet, paradoxically, black holes seem to be in thermal equilibrium with a thermal bath [11]. In other words, they have a temperature [12]. This is not allowed by the classical theories of physics, General Relativity and Thermodynamics, and Hawking found that only a quantum treatment of the fields near the black hole could explain this effect. He found that field modes of *positive* and *negative* frequency would mix when propagating through the region of curved spacetime that surrounds the hole [13]. This mixing results in spontaneous emission of particles (photons, electrons and neutrinos) that propagate away from the horizon out to infinity, where an observer would thus detect a thermal flux coming from the hole—Hawking radiation.

Many of the above statements are, to say the least, surprising. The fact that black holes emit particles by the Hawking effect is surprising. That this results from the mixing of waves with *positive* and *negative* frequency is also surprising. Yet, although the former surprise was very much of a scientific revolution (and we shall get back to this shortly), it is, after all, only one more of those "black hole surprises". The second surprise, is perhaps a bit more questioning on an intuitive level: it implies that waves, including light waves, may oscillate with positive and negative frequencies— but most people never encounter such negative frequency waves. Nevertheless, this is actually not a novelty.

Indeed, negative frequency waves are present in all field theories, but they are usually ignored because they are suggestive of redundancies in terms of information content. And yet, all fields have modes of oscillation of, both, positive and negative frequency. This is best illustrated when a field A is expressed by Fourier transform,

$$A(z, t) = \frac{1}{2\pi} \int_{-\infty}^{+\infty} d\omega \widetilde{A}(z, \omega) e^{-i\omega t}. \tag{1.1}$$

Here, the Fourier transform of the field $\widetilde{A}(z, \omega)$ has been integrated over both positive and negative frequencies ω. For a real field, the complex conjugate of the field, $A^\star$, equals the field and hence $\widetilde{A}(-\omega) = \widetilde{A}^\star(\omega)$. Accordingly, the negative component is entirely dependent on the positive component.

In fact, the apparent redundancy stated above is not always true: there exist physically realisable conditions under which the positive and negative frequency components of a field may be observed independently. Actually, they can even be made to mix. For example, this mixing has been observed in nonlinear optics, in an experiment in which energy was transferred from a wave with positive frequency to a wave with negative frequency [14]. As mentioned in the short introduction to the Hawking effect above, the motion of waves in the vicinity of a black hole is another example of these conditions under which positive and negative frequencies mix—which results in the emission of quanta.[2] If the latter effect is quantum in nature, the mixing of positive and negative frequency waves *per se*, and their very existence, has actually nothing to do with quantum physics at all.[3] As stated above, these exist in all field theories.

Field theory—waves of positive and negative frequency

I will briefly elaborate upon the existence of negative frequency waves in classical Field Theory before delving into the abstract considerations of Quantum Field Theory on curved spacetimes. To this end, I will introduce the topic of field theory via the study of a one-dimensional string, and demonstrate how negative frequencies thus arise.

At first, a string can be considered as a many-body system, a collection of point masses connected together by "springs". A continuous system with a uniform density and tension emerges when the number of point masses goes to infinity, and the distances that separate them go to zero. We begin with a collection of N points of mass m coupled together with a spring constant k such that they form a string of overall length L. Consider a string closed on itself in a circle, as a ring. Then the oscillators can be assumed to be moving about their equilibrium positions in a periodic pattern. A ring of radius considerably greater than the equilibrium separation can be treated as a linear system with periodic boundary conditions. The oscillators are constrained to vibrate along the circumference of the ring. The first and last oscillators are identical, so that if the ith oscillator is displaced from equilibrium by $\bar{\phi}_i$, the periodic boundary conditions are

$$\begin{cases} \bar{\phi}_0 = \bar{\phi}_N, \\ \frac{d\bar{\phi}_0}{dt} = \frac{d\bar{\phi}_N}{dt}. \end{cases} \tag{1.2}$$

---

[2]It is in fact in the process of articulating the latter paradigm that the former discovery was made.

[3]In the Hawking effect, quantum vacuum fluctuations in these different modes are what causes the emission, see Chap. 3.

The kinetic and potential energy (KE and PE) of the system are

$$KE = \frac{1}{2}m \sum_{i=0}^{N-1} \left(\frac{d\bar{\phi}_i}{dt}\right)^2$$

$$PE = \frac{1}{2}k \sum_{i=0}^{N-1} \left(\bar{\phi}_{i+1} - \bar{\phi}_i\right)^2. \tag{1.3}$$

In the continuum limit, the distance that separates the point masses $l \to 0$, $N \to \infty$, the length becomes $L = Nl$ and the, fixed, mass per unit length and string tension are $\mu = m/l$ and $T = kl$, respectively. Then, the displacement and energy of the string can be defined in terms of a continuous field $\bar{\phi}_i(t) = \bar{\phi}(z_i, t) \to \bar{\phi}(z, t)$, where

$$KE = \frac{1}{2}\frac{m}{l} \sum_{i=0}^{N-1} l \left(\frac{d\bar{\phi}_i}{dt}\right)^2 \to \frac{1}{2}\mu \int_0^L dz \left(\frac{\partial\bar{\phi}(z,t)}{\partial t}\right)^2$$

$$PE = \frac{1}{2}kl \sum_{i=0}^{N-1} l \left(\frac{\bar{\phi}_{i+1} - \bar{\phi}_i}{l}\right)^2 \to \frac{1}{2}T \int_0^L dz \left(\frac{\partial\bar{\phi}(z,t)}{\partial z}\right)^2. \tag{1.4}$$

The field function $\bar{\phi}(z, t)$ represents the displacement of an infinitesimal mass from its equilibrium position at z. The Lagrangian density of this continuous string is given by

$$\int_0^L dz \mathcal{L}(z, t) = \int_0^L dz \left(\frac{1}{2}\mu \left(\frac{\partial\bar{\phi}(z,t)}{\partial t}\right)^2 - \frac{1}{2}T \left(\frac{\partial\bar{\phi}(z,t)}{\partial z}\right)^2\right). \tag{1.5}$$

We introduce the wave velocity $v^2 = \frac{T}{\mu}$ and, for simplicity, we substitute $\bar{\phi} \to \sqrt{T}\bar{\phi} = \phi$. Then,

$$\mathcal{L} = \frac{1}{2} \left(\frac{1}{v^2} \left(\frac{\partial\phi}{\partial t}\right)^2 - \left(\frac{\partial\phi}{\partial z}\right)^2\right). \tag{1.6}$$

Calling upon the principle of least action allows to derive the equations of motion for the string from the Lagrangian density. Since the boundary conditions (1.2) are periodic, the boundary terms are zero. Moreover, the variation $\delta\phi$ at the initial and final time are zero. The famous Euler-Lagrange equation of motion for a continuous field is

$$\frac{1}{v^2}\frac{\partial^2\phi}{\partial t^2} - \frac{\partial^2\phi}{\partial z^2} = 0. \tag{1.7}$$

So far, we have shown how a quantity referred to as a *continuous field* emerges as the natural way to describe a system with infinitely many particles. It is the "displacement" of the dynamical system whereby $\frac{\partial\phi}{\partial t}$ is a generalized velocity and the Euler-Lagrange equation (1.7) is a wave equation. The solutions of the wave equation which satisfy the periodic boundary conditions are called the normal modes of the string. They are

$$\phi \sim e^{\pm i(k_n z - \omega_n t)}, \tag{1.8}$$

where periodicity requires $k_n = \frac{2\pi n}{L}$, $n = 0, \pm 1, \pm 2, ...$, with, respectively, k_n and ω_n the wavenumber and frequency of the wave. Inserting (1.8) in (1.7) gives the wave equation

$$\omega_n^2 = v^2 k_n^2. \tag{1.9}$$

We have arrived at a wave equation with both *positive* and *negative* frequency solutions! Let us write the states with positive frequency as

$$\phi_n(z, t) = \frac{1}{\sqrt{L}} e^{i(k_n z - \omega_n t)}, \tag{1.10}$$

in which case the negative frequency states thus have a time factor $e^{i\omega_n t}$. Since the wavenumber k_n is either positive or negative, it is convenient to express the negative frequency states as the complex conjugate of the states (1.10), $\phi_n^*(z, t)$. Thus the positive and negative frequency states are related by complex conjugation.

We could conclude this short study here, but it is actually interesting to ask ourselves the question of the normalisation of states of positive and negative frequency. We use the canonical momentum

$$\pi(z, t) = \frac{\partial \mathcal{L}}{\partial \left(\frac{\partial \phi}{\partial t} \right)} = \frac{1}{v^2} \frac{\partial \phi}{\partial t} \tag{1.11}$$

and Noether's theorem[4] to write the normalisation condition

$$\frac{i}{2} \int dz \left(\pi_1^* \phi_2 - \phi_1^* \pi_2 \right) = \frac{i}{2v^2} \int dz \left(\frac{\partial \phi_1^*}{\partial t} \phi_2 - \phi_1^* \frac{\partial \phi_2}{\partial t} \right), \tag{1.12}$$

so the states are not orthogonal in the usual sense. And yet, it can be shown that each normal mode behaves as an independent simple harmonic oscillator. Wherefrom quantisation of the field consists in quantising those oscillators.

Later in this dissertation, we shall show how scalar products similar to (1.12) can be used to define a *pseudo* norm for modes of the field—whence positive frequency states are ascribed a positive norm, and negative frequency states are ascribed a negative norm. In Field Theory, and in particular in Quantum Field Theory, it is usual to refer to modes of the field by their pseudo norm, that is, one speaks of positive- and negative-norm modes. In Optics, and in particular in Quantum Optics, the concept of pseudo-norm is not very well known, and practitioners would label the modes according to their frequency. The norm nomenclature is more general than

---

[4] See 3.2.2 for details.

the frequency one—for the latter may depend on the frame of reference, whilst the former is frame invariant. Thus, it would be generally more correct to label modes by their norm. However, in this dissertation, both the QFT and QO communities are addressed and, in an effort to use their vernacular, we will use both nomenclatures as follows: we will gradually move from considerations of the frequency of a mode to that of its norm when frame transformations come into play, and explicitly show how the norm is defined in terms of the frequency (as measured in a given frame).

In arriving at (1.8), we have not used any tools of Quantum Physics. Instead, we have merely considered a very-many-body system in the continuous limit—a field—and shown how the wave equation of this system accepted mode solutions of positive and negative frequency. Hence we have shown that the classical, real[5] field oscillates with both positive and negative frequencies. This is also true for other fields, such as the electromagnetic field. In this dissertation, we will use the second quantisation scheme to study various field theories. At various stages in what will follow, the fundamentals of field theory outlined here will be called upon to describe wave motion in General Relativity, in fluid flows, and in condensed matter systems such as optical fibres and bulk silica. We will see how fields in these media have positive and negative frequency modes of oscillation, and how these can mix, which results in spontaneous emission of quanta of this field from the vacuum (the state of lowest energy of the field).[6] The most famous instance of this effect is undoubtedly Hawking radiation—which brings us back to our earlier statement that "black holes emit a thermal flux".

Hawking radiation, the questions it raises, and the impossibility to observe it

Hawking's finding poses many questions, such as the exact origin of the flux, the fate of black holes if this outflowing energy comes at the expense of their mass, and the information content of the flux. Moreover, the very validity of the assumptions upon which Hawking's calculation is based are under question. Indeed, the effect of the mass of a black hole on the surrounding spacetime is a stretch: to an observer away from the hole, light emitted from an infalling object appears more and more redshifted as the object approaches the horizon. Actually, the redshift of light emitted from the horizon is infinite [8]. If waves from which Hawking radiation originates were to have propagated from infinity in the past through the region of the collapse and out to infinity, where the thermal flux can be observed, they will have experienced an exponential redshift from the region near the horizon to infinity. In other words, they would have had to have absurdly large frequencies in the past to be a finite-frequency, detectable, flux at late times. This is called the Trans-Planckian Problem, because it hints at some unknown physics that is at play when the wavelength of light is shorter than the Planck scale. Finally, the temperature of Hawking radiation is inversely proportional to the mass of the black hole from which it seems to escape.

---

[5]See Appendix A for further comments on this.

[6]Such an equation as (1.12) can be used as the basis for a quantum mechanics capable of describing particle production and annihilation fully: the *second quantised* form of the theory. In this scheme, negative norm modes are associated with the creation operator of the field, whilst their positive norm counterparts are associated with the annihilation operator of the field.

A very light black hole, of about 3 times the mass of the Sun, would thus radiate with a temperature 8 orders of magnitude lower than the cosmic microwave background, the Universe's own glow. Hawking radiation cannot be observed from astrophysical black holes.

Laboratory-based analogues of the event horizon

Of course, if the story had ended with this conclusion, the present dissertation would not have been written, some 43 years after Hawking's prediction. This is where the science of analogue gravity comes into play. In order to present the arguments that form the foundations of the work presented in this Thesis, let us discuss the analogy between the motion of waves on curved backgrounds and their propagation in a flowing fluid.

The prediction that a thermal flux propagates away from the vicinity of black holes was a scientific revolution—but to a rather small community of people. Indeed, in the early 1970s, the very existence of black holes was still not unequivocally established, and their physics was largely unknown, even to physicists. In a seminar that he gave at Oxford, Bill Unruh then explained that one could draw a kinematic analogy between the flow of a river toward a waterfall and the effect of black holes on spacetime. Picture a river that flows toward a waterfall: the velocity flow of the river will increase as the waterfall approaches. It may be that this increase is such that at a certain point the flow velocity equals the speed of sound, and beyond this point the flow velocity would be supersonic (and still increasing, until the water reaches the bottom of the fall). Away from the fall, because the flow velocity is subsonic, sound waves may propagate up- or downstream. Although, the closer to the fall they are emitted and the more they redshift (they shift to lower and lower frequency). A sound wave emitted *exactly* at the point at which the flow velocity of the river equals the speed of sound would not be capable of propagating upstream, against the flow, without experiencing an infinite redshift. Sound waves emitted beyond this point, down the stream, would be washed out toward the bottom of the fall, doomed. The point at which the flow velocity of the river equals the speed of sound is the sonic analogue to the event horizon: to sound waves, it is the point of non-return. It separates a region of sub- from a region of super-sonic flow of the fluid in which sound waves propagates. In analogy, the event horizon in astrophysics separates two distinct regions of spacetime: the outside region in which light waves may propagate toward or away from the central singularity, and the inside region in which wave motion is only possible toward the central singularity.

In 1981, Unruh realised that this *analogy* is not a mere *metaphor*—which renders the kinematics of waves on curved spacetimes more amenable to the intuition of physicists and a lay audience alike—but actually is *genuine*: the wave equation for certain fluids is indeed identical to the equation describing the motion of waves at the horizon, the *metric* of curved spacetime [6]. *Ergo*, some manifestations of black hole physics, namely the motion of waves in their vicinity, may be reproduced in the laboratory! Unruh called fluid black hole analogues "dumb holes" and showed that, in total analogy with their astrophysical counterparts, they should emit a thermal flux: he calculated that quantum hydrodynamical fluctuations in a moving fluid (described by a curved Lorentzian manifold) would convert into pairs of phonons at the sonic horizon—thus reviving the hopes to shed light on the Hawking emission mechanism.

Many analogue systems have been proposed and studied over the past 36 years: liquid helium [15], water waves [16, 17], sound waves in Bose-Einstein condensates [18], slow light [19, 20], electromagnetic waves in waveguides [21] or superconducting circuits [22] to name a few. The main focus of this Thesis is light in dispersive media.

Inspired by Unruh's finding [6] and the waveguide-based proposal of Unruh and Schüzhold [21], a collaboration of the groups of Leonhardt and König at St Andrews demonstrated the feasibility of creating analogue event horizons with a moving refractive index profile in dispersive optical media in 2008 [23]. An optical event horizon can be created by changing the speed of light (i.e., the refractive index of the medium of propagation) with light itself. For example, a short and intense laser pulse locally raises the refractive index of a medium by the Kerr effect: under the pulse, waves will be slowed. Hence, the profile of refractive index created by the propagating pulse effectively sets the curvature of the spacetime on which waves propagate. If light under the pulse is slowed below the pulse speed, the pulse moves superluminally, and two horizons are formed at the boundaries between sub- and superluminal propagation: light cannot enter the back of the pulse or is captured falling into the front of the pulse. In analogy with the motion of spacetime in the River Model of the black hole, the back and the front of the pulse thus act as a white hole or a black hole event horizon, respectively.[7] The authors calculated that light emitted at the horizon would be in a thermal state over a narrow band of frequencies.

As in the case of emission at the astrophysical black hole horizon, spontaneous emission from the vacuum in analogue systems results from the mixing of field modes of positive and negative frequency at the dumb hole and optical horizons. The motion of waves in laboratory systems is influenced by dispersion, which limits the extent to which waves in the medium may shift in frequency. Dispersion appears to be the analogue to transplanckian physics, with the advantage that the phenomenology is perfectly understood. Thus, the study of analogue horizons may be helpful in understanding the effect of Hawking radiation.

The observation of the generation of negative frequency waves at the horizon in water waves experiments by Rousseaux and Leonhardt in 2008 [16], and the confirmation of this effect (and of its thermal nature) by Weinfurtner and Unruh in 2011 [17], were seminal contributions to the field of analogue horizons. They clearly established that the classical effect of mixing of negative- and positive-frequency modes at the horizon is genuine, and demonstrated the need for ultra-low temperature fluid analogues to detect Hawking radiation. In 2016, Steinhauer announced having observed the entanglement of the emission of sonic waves on either sides of a black-hole horizon in a Bose-Einstein condensate (BEC) analogue [24]. Simultaneously, Rousseaux and Parentani reported on the measurement of the two-point correlation of the randomly fluctuating free surface (i.e., noise) created by the scattering of long-wavelength waves at a black-hole horizon in a water tank [25]. The findings are very

---

[7]The white hole is the time-reversed black-hole-solution to the Einstein's equations, see 2.1.2.

interesting [26] and the claims put forth by the authors have to be scrutinised. As for the water waves experiment [27], the observed correlations do not arise from quantum vacuum emission at the horizon: it is noise, a classical state, that scatters at the horizon and not quantum hydrodynamical fluctuations. Thus the observed signal results from stimulated emission at the horizon and not spontaneous emission—which is the effect ultimately sought. In the case of the BEC experiment [24], the temperature of emission at the horizon could not be properly estimated either theoretically or experimentally because the flow velocity gradient was not directly measured [28, 29]. In those experiments, the detection of individual quanta was not possible, which renders further independent measurements in different analogue systems necessary. For example, the quantum state at the output would be best characterised by a robust measurement of entanglement such as a Bell-type measurement [30]. This would allow to undoubtedly establish that spontaneous pair emission has been observed.

In that regard, optical analogues are an attractive platform that can contribute significantly to the articulation of the paradigms of analogue gravity physics and spontaneous emission from the vacuum. There exist numerous well developed theoretical frameworks for a fully quantum description of the interaction of light with matter in a dispersive medium, and the techniques developed at the crepuscule of the 20th century and dawn of the 21st century allow for precise control of the experimental parameters. In particular, the science of propagation of intense and ultrashort pulses in optical fibres and the technology of single photon counting have reached a level of refinement that enables single quantum detection at the output of an optical fibre. Optical horizons are the only analogue system that will allow for the unambiguous detection of the pair of positive- and negative frequency particles emitted at the horizon—a signature of the Hawking effect.

Content and structure of the dissertation

This dissertation presents the theoretical and experimental study of the scattering of light on transient inhomogeneities in highly dispersive media. Spontaneous, and stimulated, emission from the vacuum in various systems, and the kinematics and mathematical arguments that support the analogy between laboratory systems and astrophysical black holes are the central problem around which the various chapters are organised.

In Chap. 2, we begin with an attempt to measure the time and position at which a moving clock ticks. In a first section, we introduce in this way the concepts of events and relativity of measurements of events, which are fundamental concepts in the theories of Special and General Relativity. The transition from Special to General Relativity is then achieved in an elementary fashion by studying the curvature of spacetime. We study conditions of extreme curvature, and in particular the case in which there is a spacetime singularity surrounded by an event horizon—a black hole. We proceed to elaborate upon the effects of the curvature of spacetime on waves—namely one way motion inside the horizon and infinite frequency shift upon propagation from the horizon out to infinity. This allows us to identify key physical phenomena characteristic of wave motion at the horizon, wherefrom we introduce the

River Model of the black hole and draw the analogy to fluid mechanics by following the steps laid out by Unruh.

The formalism is transferred to optics in a second section, when we study the motion of waves in optical fibres. We explain how a solitary wave may form in a fibre as a result of the interplay between nonlinear and dispersive effects in the medium. In particular, we show how the refractive index is increased under a *soliton*, and how other waves will experience this transient change in the medium properties upon interaction and slow down and shift in frequency. We invoke kinematics arguments of the River Model of the black hole to show how the propagation of a soliton in the fibre effectively creates a moving horizon. This leads us to considering the frequency of oscillation of light waves in the fibre, which yields a discussion of the first observation of the energy transfer from a positive frequency to a negative frequency wave. Finally, we use a toy-model for a field theory of light-matter interaction to show the full mathematical analogy between waves in an optical medium and waves at the astrophysical horizon.

In the first section of Chap. 3, we look back upon black hole physics and demonstrate how they resemble thermodynamic objects, which leads us to a paradox: how could they possibly emit heat since nothing can propagate out of them from beyond the horizon out to infinity? This failure of classical physics is attended to by means of a quantum study of a scalar field on a curved background. We use second quantisation to derive Hawking's seminal result that black holes emit radiation with a thermal spectrum, and discuss the implications of, and questions related to, this finding. A similar problem is considered in the second section of Chap. 3, where we quantise a field theory for light-matter interaction in a dispersive medium [31, 32]. A careful study of the dispersion relation allows to draw kinematics analogies with the case of wave motion on a curved background, and various such configurations are found to be simultaneously realised at a moving interface in the refractive index of the medium. We then show how the mixing of modes with positive and negative frequencies at this interface leads to spontaneous emission from the vacuum and discuss the implications of this effect.

Chapter 4 is dedicated to the presentation of the numerical results of this Thesis: spectra of spontaneous emission at a moving interface in a dispersive medium. The first section presents the analytical method developed to model the interactions of light with matter considered in Chap. 3. In particular, the unique algebra necessary to account for the various configurations of modes at the interface in the refractive index is thoroughly explicated, and its use exemplified. We then show how this method may be implemented to compute spectra of emission as they can be observed in the laboratory frame. Numerically computed spectra for bulk silica are presented in a second section, where we find that the strongest emission is in a mode that has a *negative* optical frequency.

The experiment conducted to observe the effects of waves of positive and negative frequency scattering at a soliton is presented in the last chapter of the dissertation, Chap. 5. We begin with a derivation that shows that a coherent, continuous wave, probe of positive frequency would transfer energy to a wave with negative frequency (observable in the UV) upon scattering at the soliton. This parametric amplification of waves with opposite-sign frequency has never been observed before in optics. We discuss the place of the experiment in the field in the light of other studies of stimulated emission at the horizon in water waves setups [16, 17, 25] and its relation to the nonlinear optics experiment [14] that established the reality of negative frequency waves in optics. After explaining the setup, we characterise the ability of our apparatus to resolve and detect a negative frequency signal. In the next section, we present the classical effect of horizons on waves: the shift of frequency experienced by waves impinging on the horizon. We explain how this positive frequency to positive frequency wave scattering effect is a signature of horizon physics. Advances toward observing the negative frequency signal are then discussed via a study of the signal to noise ratio in the UV. In particular, we report the observation of a peak at 247 nm. We ponder upon the origin of this signal and the possibility of this peak being related to the signal sought. We then conclude with a discussion of future developments of this experiment and considerations of the route toward the detection of spontaneous emission of light quanta in an optical fibre.

References

1. A. Einstein, Die feldgleichungen der gravitation, in *Sitzungsberichte der Preussischen Akademie der Wissenschaften zu Berlin* (1915), pp. 844–847
2. A. Einstein, Die grundlage der allgemeinen relativitatstheorie. Ann. der Phys. **354**(7), 769–822 (1916)
3. J.C. Maxwell, On physical lines of force. Phil. Mag. **11**, 11611–175; 281–291; 338–348 (1861)
4. J.C. Maxwell, On physical lines of force. Phil. Mag. **12**(12–24), 85–95 (1862)
5. J.C. Maxwell, *A Treatise on Electricity and Magnetism.* (Clarendon press edition, 1873)
6. W.G. Unruh, Experimental black-hole evaporation? Phys. Rev. Lett. **46**(21), 1351–1353 (1981)
7. S.W. Hawking, Black hole explosions? Nature **248**(5443), 30–31 (1974)
8. C.W. Misner, K.S. Thorne, J.A. Wheeler, *Gravitation.* (W.H. Freeman, San Francisco, 1973)
9. R. Penrose, Gravitational collapse and space-time singularities. Phys. Rev. Lett. **14**(3), 57–59 (1965)
10. LIGO scientific collaboration and virgo collaboration. Observation of gravitational waves from a binary black hole merger. Phys. Rev. Lett. **116**, 061102 (2016)
11. J.M. Bardeen, B. Carter, S.W. Hawking, The four laws of black hole mechanics. Commun. Math. Phys. **31**(2), 161–170 (1973)
12. J.D. Bekenstein, Black holes and entropy. Phys. Rev. D **7**(8), 2333–2346 (1973)
13. S.W. Hawking, Particle creation by black holes. Commun. Math. Phys. **43**(3), 199–220 (1975)
14. E. Rubino, J. McLenaghan, S.C. Kehr, F. Belgiorno, D. Townsend, S. Rohr, C.E. Kuklewicz, U. Leonhardt, F. König, D. Faccio, Negative-frequency resonant radiation. Phys. Rev. Lett. **108**(25) (2012)
15. G.E. Volovik. *The Universe in a Helium Droplet.* Number 117 in International series of monographs on physics (Oxford University Press, Oxford, 2009) OCLC: 636215451
16. G. Rousseaux, C. Mathis, P. Massa, T.G. Philbin, U. Leonhardt, Observation of negative-frequency waves in a water tank: a classical analogue to the Hawking effect? New J. Phys. **10**(5), 053015 (2008)
17. S. Weinfurtner, E.W. Tedford, M.C.J. Penrice, W.G. Unruh, G.A. Lawrence, Measurement of stimulated hawking emission in an analogue system. Phys. Rev. Lett. **106**(2) (2011)
18. O. Lahav, A. Itah, A. Blumkin, C. Gordon, S. Rinott, A. Zayats, J. Steinhauer, Realization of a sonic black hole analog in a bose-einstein condensate. Phys. Rev. Lett. **105**(24) (2010)

19. U. Leonhardt, A laboratory analogue of the event horizon using slow light in an atomic medium. Nature **415**(6870), 406–409 (2002)
20. R. Schützhold, G. Plunien, G. Soff, Dielectric black hole analogs. Phys. Rev. Lett. **88**, 061101 (2002)
21. R. Schützhold, W.G. Unruh, Hawking radiation in an electromagnetic waveguide? Phys. Rev. Lett. **95**, 031301 (2005)
22. P.D. Nation, M.P. Blencowe, A.J. Rimberg, E. Buks, Analogue hawking radiation in a dc-squid array transmission line. Phys. Rev. Lett. **103**, 087004 (2009)
23. T.G. Philbin, C. Kuklewicz, S. Robertson, S. Hill, F. König, U. Leonhardt, Fiber-optical analog of the event horizon. Science **319**(5868), 1367–1370 (2008)
24. J. Steinhauer, Observation of quantum hawking radiation and its entanglement in an analogue black hole. Nat. Phys. **12**(10), 959–965 (2016)
25. L.-P. Euvé, F. Michel, R. Parentani, T.G. Philbin, G. Rousseaux, Observation of noise correlated by the hawking effect in a water tank. Phys. Rev. Lett. **117**, 121301 (2016)
26. W.G. Unruh, Experimental black hole evaporation. Phys. Today (2016)
27. E. Léo-Paul, G. Rousseaux, Génération non-linéaire d'harmoniques après une conversion linéaire en interaction houle-courant, in *XIVèmes Journées Nationales Génie Côtier Génie Civil*, ed. by D. Levacher, M. Sanchez et, V. Rey (Editions Paralia CFL, Nantes, 2016), pp. 181–190
28. F. Michel, J.-F. Coupechoux, R. Parentani, Phonon spectrum and correlations in a transonic flow of an atomic bose gas. Phys. Rev. D **94**(8) (2016)
29. U. Leonhardt, Questioning the recent observation of quantum Hawking radiation (2016), arXiv:1609.03803
30. A. Finke, P. Jain, S. Weinfurtner, On the observation of nonclassical excitations in bose einstein condensates. New J. Phys. **18**(11), 113017 (2016)
31. S. Finazzi, I. Carusotto, Quantum vacuum emission in a nonlinear optical medium illuminated by a strong laser pulse. Phys. Rev. A **87**(2) (2013)
32. M. Jacquet, F. König, Quantum vacuum emission from a refractive-index front. Phys. Rev. A **92**(2) (2015)

Chapter 2
Theory of Spacetime Curvature in Optical Fibres

From astrophysics to the laboratory, and more precisely to optical fibre systems, this chapter will present the fundamentals of the science of analogue spacetimes realisations. Leaving the concepts of quantum field theory—in curved spacetime and for light-matter interaction—that describe the spontaneous creation of light from the vacuum to a later chapter, here we focus on classical physics in its most modern form: starting from Special Relativity and moving on to introduce concepts of General Relativity, we will show how a black hole influences the fabric of the universe. Considering the generative idea of analogue spacetimes, the flow of the above-mentioned fabric and how it can be recreated in laboratory systems will then lead us to investigate our experiment: light in optical fibres.

2.1 From Astrophysics to the Laboratory

2.1.1 Curvature of Spacetime

The theories of relativity, Special and General Relativity, have been used since the first half of the twentieth century to explore the boundaries of Nature. In this section we wish to gain an understanding of physics in the vicinity of black holes—General Relativity, the Theory of Gravitation that describes matter and motion near massive objects. Discussing the key concepts of Special Relativity, the theory of the very fast, will provide us with the principles essential to examine spacetime curvature as described by General Relativity. We will see how spacetime curves in the vicinity of a black hole and what metric best describes this phenomenon.

© Springer International Publishing AG, part of Springer Nature 2018
M. J. Jacquet, *Negative Frequency at the Horizon*, Springer Theses,
https://doi.org/10.1007/978-3-319-91071-0_2

2.1.1.1   Special Relativity

Our journey with Relativity begins with a clock that ticks and moves past a point
in a particular inertial frame. We seek to measure the distance s and time t between
two ticks. t is referred to as the time separation between two subsequent events,
and s is called the space separation. Our common experience, for example with the
siren of an ambulance rushing by us in the street, tells us that these space and time
separation are frame dependent quantities. Yet, all inertial observers agree on the
time as measured by the clock between two events in the frame of the clock. This is
the *proper time τ*,

$$\tau^2 = t^2 - s^2 \tag{2.1}$$

Because $t > s$, it is also referred to as the timelike spacetime interval. τ is invariant, that is frame independent. The above equation recalls the Pythagorean theorem
($\tau^2 = t^2 + s^2$) which gives the distance between two points in Euclidean space. Similarly, this metric gives the separation between any two events (for which $t > s$) in
spacetime. It provides all information about the (non quantum) features of spacetime
and can be extended to predict trajectories. One can also find the proper distance, or
spacelike spacetime interval, between two events:

$$\sigma^2 = s^2 - t^2. \tag{2.2}$$

A spacetime interval is then defined as the combination of the spacelike and timelike
spacetime intervals of the metric.

Special Relativity is valid within the inertial frame, that is a flat region of spacetime. In terms of the mathematics of Relativity, this is the realms of particular
pseudo-Riemannian metrics. Take g, a symmetric covariant 2-tensor field (a pseudo-
Riemannian metric on a manifold V), defined such that

$$(g_{ij})^{-1} = (g)^{ij}. \tag{2.3}$$

In a moving frame, it is written as

$$g = g_{ij}\theta^i\theta^j, \tag{2.4}$$

with $\theta^i = \{t', x', y', z'\}$ the coordinates in the moving frame, and in a laboratory
(natural) frame as

$$g = g_{ij}dx^i dx^j \tag{2.5}$$

with $x^i = \{t, x, y, z\}$ the coordinates in the natural frame. By definition, g is a non-
degenerate quadratic form: it can be written as a sum of independent real linear forms
of dx^i in the moving frame

$$g_{ij}dx^i dx^j \equiv \sum_i \epsilon_i(\theta^i)^2, \ \epsilon_i = \pm 1. \tag{2.6}$$

The number of $\epsilon_i = 1$ or -1 is the signature of the metric. For all point $x \in V$, g defines a scalar product between two four-vectors of the vector space

$$(v, w) := g_x(v, w) = g_{ij}(x)v^i w^j. \tag{2.7}$$

A flat spacetime is a manifold isometric with a pseudo-Euclidean space (that is Re^n) with metric

$$g_{ij}dx^i dx^j \equiv \sum_i \epsilon_i (dx^i)^2, \ \epsilon_i = \pm 1. \tag{2.8}$$

And a spacetime is locally flat if the manifold V is locally isometric to a flat space. Flat spacetime is best and most often described by the Minkowski metric, a flat metric on Re^{n+1}:

$$g \equiv -(dx^0)^2 + \sum_{i=1}^{n} (dx^i)^2 \tag{2.9}$$

which has a $(-++\cdots+)$ signature in the Misner–Thorne–Wheeler convention [1]. In the natural frame (Cartesian coordinates) it is the symmetric, position dependent, matrix

$$g = \begin{pmatrix} -1 & 0 & 0 & 0 \\ 0 & 1 & 0 & 0 \\ 0 & 0 & 1 & 0 \\ 0 & 0 & 0 & 1 \end{pmatrix}. \tag{2.10}$$

An important concept in Relativity is that of geodesics, the path that a particle which would not be accelerating follows. It is said of a geodesic that it is a locally separation-extremising curve. In other words, geodesics are curves that locally give the shortest distance between two events. To define a geodesic, one has to call on the principle of causality: because the Minkowski metric is a Lorentzian metric, it is time oriented, and so is the manifold (flat spacetime, Re^{n+1}) on which it is defined. The length of a causal curve γ joining x_a to x_b on V is

$$l := \int_a^b L d\lambda, \tag{2.11}$$

with the Lagrangian

$$L := g_{\alpha\beta}(x(\lambda))\dot{x}^\alpha \dot{x}^\beta, \ \dot{x}^\alpha := \frac{dx^\alpha}{d\lambda} \tag{2.12}$$

with parameters $\lambda = a$ and $\lambda = b$ (λ is the path length). Then a geodesic joining x_a and x_b is defined as a solution of the Euler equation for this Lagrangian:

$$\frac{d}{d\lambda} \frac{\partial L}{\partial \dot{x}^\alpha} - \frac{\partial L}{\partial x^\alpha} = 0. \tag{2.13}$$

On flat spacetime, the distance between nearby parallel geodesics is constant. The acceleration of distance between nearby geodesics is an indicator of a curved spacetime. In the cosmos, most regions of spacetime are flat over only a limited range of space and time. If a pair of free test particles experiences a relativistic acceleration with respect to one another (relativistic tidal forces), then spacetime is not flat but curved. When this is the case, the prevailing theory of Physics is no longer Special Relativity but General Relativity. Einstein's genius was to realise that the curvature of spacetime, the rate of acceleration of distance between nearby geodesics, was identical to relativistic tidal forces.

2.1.1.2 General Relativity

Newton would appeal to the principle of material indifference to express the idea of the invariance of physical laws and phenomena upon the reference frame in which they are expressed. Einstein extended this principle by relying on tensor fields as objects for physical laws. Since those are intrinsic objects on a manifold, they are represented by their frame specific components but pass from one frame to the other via general laws. The extension to Newton's principle, *general covariance*, and the Newton–Galileo equivalence principle (the independence of the acceleration of a body on its mass under gravity) inspired Einstein to use Lorentzian differential geometry and ushered-in the theory of General Relativity.

Replacing the Minkowski spacetime of Special Relativity by a general Lorentzian manifold and appealing to the Newton equivalence principle between inertial and gravitational masses allowed Einstein to invent General Relativity. According to this theory, a four-dimensional curved Lorentzian manifold unites space and time (which were previously considered as a priori given structures). The metric of this spacetime is linked with its curvature and governs its causality structure. Both in Special and General Relativity, the basic observable quantity is the length of a timelike curve as a measure of the proper time between two events. A massive object in free fall follows the timelike geodesics of the metric: their equations of motion are independent of their mass. Light rays are null geodesics in General Relativity, i.e. the trajectories of particles with zero rest mass.

There exist various forms of the Einstein field equations [2, 3], for the purpose of the current discussion it suffices to appeal to the Einstein equation in vacuum. This states that the Lorentzian metric g must satisfy

$$Ricci(g) = 0, \tag{2.14}$$

with *Ricci* the Ricci curvature tensor, a special Riemannian tensor that describes the curvature of spacetime (see [4] for more details). It is zero because the stress-energy tensor in space outside a star is zero, we are in vacuum. This equation for the gravitational field means that, in General Relativity, gravity is identified with the curvature of spacetime—a massive object will deform spacetime.

Soon after the publication of Einstein's theory of General Relativity, Scwharzschild was the first to construct an exact solution to the Einstein's equation in vacuum (2.14). His solution models the gravitational field outside spherically symmetric isolated bodies, such as the Sun or Earth, or a proton.

2.1.2 The Schwarzschild Spacetime

2.1.2.1 Schwarzschild Metric

Written in standard polar coordinates (t, r, θ, ϕ), the smooth spherically symmetric metric solution to Einstein's field equation (2.14) by Schwarzschild is [5]

$$g_{Schw} = -\left(1 - \frac{2m}{r}\right)(dt)^2 + \left(1 - \frac{2m}{r}\right)^{-1}(dr)^2 + r^2\left((d\theta)^2 + \sin^2\theta(d\phi)^2\right).$$

(2.15)

As for the Minkowski metric, we express the Schwarzschild metric in matrix form as

$$g = \begin{pmatrix} -(1 - \frac{2m}{r}) & 0 & 0 & 0 \\ 0 & (1 - \frac{2m}{r})^{-1} & 0 & 0 \\ 0 & 0 & r^2 & 0 \\ 0 & 0 & 0 & r^2 \sin\theta \end{pmatrix},$$

(2.16)

where t is the time as measured at infinity, and m is a constant that has implications on the nature of this metric. Indeed,

- when $m = 0$, the Schwarzschild metric is identical to the Minkowski metric of flat spacetime,
- when $m \neq 0$, the metric is singular for $r = 0$ and has a coordinate singularity for $r = 2m$. Clearly, the sign of the constant $m \neq 0$ is very important in determining the properties of the metric: a metric where m would be negative has been given no physical interpretation to date. On the other hand, for positive m, the Schwarzschild metric is a regular Lorentzian metric with t timelike and r spacelike $(r > 2m)$.

For $r < 2m$, $r \neq 0$, the metric is a regular Lorentzian one, but the time- and space-like character of the t and r coordinates interchanges with the case where the metric describes a flat spacetime. The case of $r = 2m$ is peculiar and deserves a discussion of its own. $1 - \frac{2m}{r}$ vanishes for $r = 2m$, implying that the Schwarzschild metric in standard coordinates is singular there: g_{00} vanishes and g_{rr} becomes infinite—the metric is not a smooth Lorentzian metric for $r = 2m$. Together with the change in time- and space-like character of the t and r coordinates for $r < 2m$, $r \neq 0$, this hints that $r = 2m$ is not a genuine singularity of Schwarzschild spacetime: it only appears to be in the (t, r, θ, ϕ) coordinates, which are unsuited to the region $r \leq 2m$. Thus one cannot make physical predictions from the Schwarzschild metric at $r = 2m$ and needs to use another coordinate system. This apparent singularity in the Schwarzschild metric is important. Indeed, no classical signal (i.e., one not due to a quantum effect) can escape from the regions $r < 2m$: the hypersurface Re $\times \{r = 2m\}$ is called an *event horizon*. A spacetime with a source of radius $a < 2m$ is called a black hole.

Let us express the Schwarzschild metric as Eddington [6], Lemaître [7] and Finkelstein [8] suggest to do. We perform a change of coordinates for $r < 2m$ from the canonical Schwarzschild time t to the retarded time v,

$$v = t + r + 2m \log(\frac{r}{2m} - 1), \tag{2.17}$$

(note that this change of coordinates from (t, r, θ, ϕ) to (v, r, θ, ϕ) is singular for $r = 2m$) to express the Schwarzschild metric in the Eddington–Finkelstein (EF) form:

$$g_{EF} = -\left(1 - \frac{2m}{r}\right)(dv)^2 + 2drdv + r^2\left(\sin^2\theta(d\phi)^2 + (d\theta)^2\right). \tag{2.18}$$

This equation defines a vacuum Einsteinian spacetime referred to as the EF black hole. As $g^{rr} = 0$ for $r = 2m$, the submanifold $r = 2m$ is a null surface. Imposing $\theta = $ constant, $\phi = $ constant, we find two families of radial light rays: one represented by straight coordinate lines $v = $ constant, and the other given by

$$-\left(1 - \frac{2m}{r}\right)dv + 2dr = 0 \Rightarrow dv = \left(-\frac{4m}{2m - r} + 2\right)dr \tag{2.19}$$

which integrates for $r < 2m$ to

$$v = 2r + 4m \log\left(|2m - r|\right) + \text{constant}. \tag{2.20}$$

In the domain $r < 2m$, the EF metric can be expressed in the Shwarzschild form via $t = v - r - 2m \ln\left(\frac{r}{2m} - 1\right)$ and reads

$$g_{EF} = \left(\frac{2m}{r} - 1\right)(dt)^2 - \left(\frac{2m}{r} - 1\right)^{-1}(dr)^2 + r^2\left((d\theta)^2 + \sin^2\theta(d\phi)^2\right) \tag{2.21}$$

with t spacelike and r timelike! The metric is not static anymore. The EF metric is genuinely singular for $r = 0$. This is in general interpreted as a spacelike 3-surface, or hypersurface. An EF spacetime is called a black hole because no future light ray issuing from a point where $r < 2m$ crosses the event horizon $r = 2m$ (see Fig. 2.1).

At this stage, we can digress shortly and remark that, upon time-reversal, there exists another extension of Schwarzschild spacetime:

$$g_{WH} = -\left(1 - \frac{2m}{r}\right)(dv)^2 - 2drdv + r^2\left(\sin^2\theta(d\phi)^2 + (d\theta)^2\right). \tag{2.22}$$

Fig. 2.1 The history of a spherically symmetrical star collapsing to form a black hole. The wavy arrows at each point indicate how an ingoing or outgoing light flash propagates. Figure and caption from [9]. Spacetime around a black hole is curved, so the worldlines of the light flashes are directed towards the 'inside'. At a characteristic distance from the singularity ($r = r_s$, the Schwarzschild radius), these worldlines are so tilted that they become vertical in the diagram. This form a surface (a cylinder of outer surface the vertical dashed line) called the event horizon

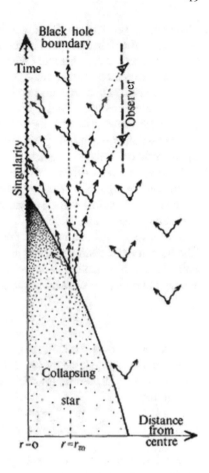

This is a white hole, nothing can penetrate into it. Indeed, every outgoing radial null geodesics in Schwarzschild spacetime emanates from the white hole.

2.1.2.2 Event Horizon

In relativity, the velocity of light in vacuum c is constant, invariant. Yet the curvature of spacetime, the dependence of the measure of proper time on the observer, causes light in a gravitational field to redshift, and time to dilate.

It can be shown (see [4]) that, in a Schwarzschild spacetime where both an object A and the observer O are at rest, the period of the radiation when emitted from the object T_A is related to the period upon observation T_O ($r_A < r_O$) by

$$T_O = \sqrt{1 - \frac{2m}{r_O}} \sqrt{(1 - \frac{2m}{r_A})^{-1}} T_A. \tag{2.23}$$

Hence the observed period is larger than the emitted period, and an observer would observe a redshift in the spectral lines of the atom radiation. In the limit of $\frac{m}{r_O}$ small, we see that, to a faraway observer (Wheeler's bookkeeper), a signal emitted from $r_A = 2m$ (source located at the horizon) would have infinite wavelength. This is called the infinite-redshift effect and has implications for the validity of the Hawking radiation temperature derivation. This infinite-redshift effect gives full physical meaning to the common saying that "nothing can escape a black hole": light emitted from the horizon would have infinite period upon observation, or, symmetrically, zero period upon emission. This is not a signal, it cannot be detected [10].

The dependence on the observer of the measure of the proper time can be illustrated by computing the radial velocity that must be applied to the object A (with respect to the observer O at rest in the Schwarzschild metric) for it to escape the gravitational attraction—the escape velocity.

As above, r_O is the coordinate of the static observer, and r_A is that of the test object. The trajectory of bodies of small size and mass in the gravitational field of a black hole are timelike geodesics of the Schwarzschild spacetime with mass $m < r_O/2$. We denote by ds the element of proper time on a timelike curve,

$$ds^2 \equiv -g_{\alpha\beta}dx^\alpha dx^\beta. \tag{2.24}$$

Thus the proper-time initial velocity is $\dot{r}_O = dr/ds(O)$, and the variation of the time parameter is denoted by $\dot{t} = dt/ds$. Our object, which is supposed to be in free motion after its departure, thus follows radial geodesic curve and hence satisfies the equation [4]

$$\left(1 - \frac{2m}{r}\right)\dot{t} = E = \left(1 - \frac{2m}{r_O}\right)\dot{t}_O, \tag{2.25}$$

where E is a constant and r_O is the r coordinate of the static observer. Note that for the sake of the present computation, we consider a Schwarzschild spacetime with mass $m < r_O/2$. As a result of the Definition 2.24,

$$1 = \left(1 - \frac{2m}{r}\right)\dot{t}^2 - \left(1 - \frac{2m}{r}\right)^{-1}\dot{r}^2. \tag{2.26}$$

Hence the differential equation for the parameter r

$$\dot{r}^2 = E^2 - 1 + \frac{2m}{r}. \tag{2.27}$$

Clearly, the maximum of r is attained for $\dot{r} = 0$ (a reversal of velocity implies that the object turns back at this point). This is for

$$r_M = \frac{2m}{1 - E^2}. \tag{2.28}$$

Obviously, the parameter r can only maximise to r_M if $E < 1$. The escape velocity, corresponding to an infinite value of r_M corresponds to $E = 1$. By (2.27), this velocity is

$$\dot{r}_O^2 = \frac{2m}{r_O}.$$

(2.29)

For the observer at rest, the relativistic escape velocity β is given by the ratio of the radial space and time components V^1 and V^0 of the velocity vector in the proper frame of this observer (of components $(\dot{r}_O, \dot{t}_O)$) in the natural frame of the coordinates t, r). The proper frame of the observer is

$$r'^0 = \sqrt{1 - \frac{2m}{r_O}} dt, \quad r'^1 = \left(1 - \frac{2m}{r_O}\right)^{-\frac{1}{2}} dr.$$

(2.30)

The velocity vector has components $(\dot{r}_O, \dot{t}_O)$ in the natural frame of the coordinates t, r. Therefore,

$$V^0 = \sqrt{1 - \frac{2m}{r_O}} \dot{t}_O, \quad V^1 = \sqrt{\left(1 - \frac{2m}{r_O}\right)^{-1}} \dot{r}_O.$$

(2.31)

And, finally, when $E = 1$, plugging (2.26) and (2.28) and rearranging yields

$$\beta =: \frac{V^1}{V^0} = \sqrt{\frac{2m}{r_O}}.$$

(2.32)

This escape velocity tends towards the velocity of light when $r_O \to 2m$.

If, now, the velocity of the object is less than the escape velocity, according to the above calculation, the time it takes it to fall back to its departure point is twice the parameter time it takes to attain r_M. We calculate the proper time this takes using

$$s_A = \int_0^{t_M} \frac{ds}{dt} dt = \int_0^{t_M} E^{-1} \left(1 - \frac{2m}{r}\right) dt = \int_0^{t_M} \frac{1 - \frac{2m}{r}}{1 - \frac{2m}{r_O}} \frac{ds}{dt}(0) dt,$$

(2.33)

with the change of variables

$$[\frac{ds}{dt}(0)]^2 = \left(1 - \frac{2m}{r_O}\right) - \left(1 - \frac{2m}{r_O}\right)^{-1} [\frac{dr}{dt}(0)]^2.$$

(2.34)

We perform yet another change of variables and set $(dr/dt)(0) = v$ (and remember that m/r_O is small) to obtain

$$s_A \cong \int_0^{t_M} \left(1 - \frac{2m}{r} + \frac{2m}{r_O} + \frac{mv^2}{r_O}\right) dt.$$

(2.35)

The proper time as measured by the observer sitting at r_O between the departure and return is

$$s_O = \int_0^{t_M} \sqrt{1 - \frac{2m}{r_O}} \, dt \cong \int_0^{t_M} \left(1 - \frac{m}{r_O} \right) dt. \tag{2.36}$$

This proper time differs from that measured in the frame of the object: the delay (within our approximation) is

$$s_A - s_O \cong \int_0^{t_M} \left(-\frac{2m}{r} + \frac{m}{r_O} + \frac{mv^2}{r_O} \right) dt. \tag{2.37}$$

Since $r > r_O$, this delay is greater than zero. The above derivation shows that, when spacetime is curved, the proper time measure depends on the observer frame. This is a significant departure from Special Relativity. Light propagating away from a black hole will, because of the steep curvature of spacetime, significantly redshift.

In an inertial frame, no signal can move in any direction faster than light: this means that the forward light-cone contains all possible worldlines for a passing particle. Since worldlines can run through a horizon only in the radial inwards direction, the horizon effectively causally disconnects the inside region from the outside region of the black hole. As we have seen earlier, no future light ray originating from inside the horizon of a Schwarzschild black hole can escape to outer space. In addition, in the inner region of the black hole, where $r < 2$ m, the radius r decreases inexorably. This means that motion inside the horizon is possible in only one way, towards the central singularity. Whatever falls on the black hole and crosses the event horizon is therefore doomed and feeds our earlier conclusion about the blackness of black holes.

In conclusion, we have seen that the Schwarzschild metric, solution to Einstein's equation for the gravitational field in vacuum, can describe situations of extreme curvature. These gravitational collapses have been given the name of black holes because they deform spacetime in such a way that anything coming too close is doomed to fall on their central singularity. The boundary between the inner and outer regions of a black hole is referred to as the event horizon. We have shown that it causally disconnects worldlines, even those of light rays, and that a signal emitted from this hypersurface would be infinitely redshifted before it reaches a far away observer (meaning that no signal can travel outwards from the event horizon).

2.1.3 Laboratory Event Horizons

2.1.3.1 The River Model of Flowing Spacetime

Let us have a closer look at the EF metric (2.18), by writing it in yet a different form, as Painlevé [11] and Gullstrand did [12]. The present form has the advantage of offering an intuitive interpretation of the distortion of spacetime in the vicinity of a black hole. It reads [13]

$$g_{PG} = -dt_{PG}^2 + (dr + \beta dt_{PG})^2 + r^2 \left((d\theta)^2 + \sin^2 \theta (d\phi)^2\right), \qquad (2.38)$$

with $\beta = \sqrt{\frac{2m}{r_o}}$ the Newtonian escape velocity (see Eq. 2.32, and note that r_O is the spatial tortoise coordinate of the static observer) and t_{PG} the proper time. In this form, the metric describes ordinary flat space. What is interesting is that space itself is flowing radially inwards at velocity β. At the horizon $r = r_{Schw}$, $\beta = c$ (equals 1 in the metrics units). This metric is the fundamental brick of the River Model of black holes.

The River Model has the same features as the original Schwarszchild metric. In particular, an object that falls through the horizon appears redshifted to an outside observer, and frozen at the horizon: t_{PG} increases so, as the object approaches the horizon, it takes an ever increasing time for the light it radiates to progress backwards against the infalling space and finally reach the observer. What is interesting is that the equations governing the propagation of sound waves in an inviscid, barotropic, irrotational fluid are identical to those for a massless scalar field propagating in a General Relativity metric [14]. In other words, such a fluid can be used to reproduce the physics of black hole event horizons. Unruh showed in 1981 that sound horizons (of what he calls 'dumb holes') emit Hawking Radiation and ushered-in the field of laboratory analogues to event horizons realisation.

2.1.3.2 Intuition of Dumb Holes

When a fluid flows faster than the speed of sound somewhere in an inhomogeneous flow, a sonic horizon appears. Since the speed of sound is the maximal speed for excitations of the fluid, the horizon separates two regions of fluid flow, a supersonic region from a subsonic region. By analogy these are identified with the inner and outer regions, respectively, of a black hole. The fluid is a self contained quantum system: its degrees of freedom are conserved. This means that, in the presence of a sonic horizon, waves incoming on the horizon (incoming modes) will convert into outgoing modes (outgoing waves). But waves cannot propagate from inside to outside the horizon of a black hole, and neither can they in the case of a sonic horizon. Yet, in the case of the fluid analogue, outgoing modes come from incoming degrees of freedom that 'turned backwards' at the horizon—an example of a general phenomenon that occurs for dispersive waves in an inhomogeneous, dispersive medium.

Sound waves, in the present context, are waves propagating on the fluid (this could be water) surface, with gravity and surface tension as the restoring forces—because of the interplay between those forces, a fluid with a free surface is considered to be a dispersive medium. The frequency dispersion of sound waves implies that waves of different wavelengths travel at different phase velocities.[1] Propagating sound waves of non-zero amplitude only can exist when the angular frequency ω and the wave number $k = \frac{2\pi}{\lambda}$ (λ the wavelength) satisfy a functional relationship known as the dispersion relation, of the form $\omega^2 = gk\tanh(kh)$, where g is the acceleration by gravity and h the depth of the water. In a co-moving frame at rest with a fluid flowing at velocity u, the dispersion relation becomes

$$(\omega - uk)^2 = gk\tanh(kh). \tag{2.39}$$

Because of dispersion, the phase velocity $v_p = \omega/k$ and group velocity $v_g = \partial\omega/\partial k$ are different: i.e., in water, the phase of sound waves will propagate at different speed from the wave packets. These velocities clearly are frequency-dependent. Or, conversely, a sound wave whose group velocity changes as it propagates in the inhomogeneous, dispersive medium, will experience a shift in frequency—and 'turn backwards' at the horizon, as mentioned above. Let us now explain this effect.

In a dispersive fluid, sound waves whose group-velocity is lower than the fluid flow-velocity will experience a black-hole horizon. Because of the frequency-dependence of the group-velocity, this horizon is not a sharp interface (unlike the absolute event horizon of astrophysical black holes). Each incoming sound wave actually experiences a continuous reversal of its group velocity as it approaches the point where the fluid flows faster than sound can. This continuous reversal of the group velocity results from a smooth evolution from one branch of the dispersion relation (2.39) of the medium to another. The phenomenology of group-velocity reversal and frequency shift at the horizon is as follows: an incoming, high frequency mode is dragged toward the horizon by the faster fluid flow. As it moves towards the horizon, the wave-vector k of the mode decreases and its group velocity increases, eventually reaching and exceeding the flow velocity. The incoming wave packet then begins converting into outgoing modes propagating back out away from, and through, the horizon, into the *outside* and *inside* region of the dumb hole, respectively. Indeed, in a dispersive medium, mode conversion does not imply that energy is transferred from one incoming mode to one outgoing mode only. On the contrary, one incoming mode can transfer energy to all of the outgoing modes upon scattering at the horizon. And, symmetrically, the energy in an outgoing mode can be contributed by all the incoming modes. Because of dispersion, outgoing modes will have different group velocity than the incoming mode. For example, under dispersion, the change in wave vector of the the 'reflected' mode is accompanied by a frequency shift, to lower frequencies: a redshift. That is, the outgoing mode is redshifted with respect to the incoming mode.

A peculiarity of inhomogeneous, dispersive media, is that they support modes with both positive and negative frequencies (as can be seen already from Eq. 2.39). Like the scattering of modes at the horizon, this surprising feature will be detailed later in this dissertation (see Sects. 2.2.2, 3.1.2 and 3.2.2 for example). Unruh showed that

---

[1] This is a linear effect (contrarily to amplitude dispersion whereby waves or larger amplitude have a different phase velocity from small-amplitude waves).

energy is converted from the positive and negative branch of the dispersion relation of the fluid in the amount predicted by the Hawking emission theory.

Over the course of the present discussion, the focus has swiftly shifted from the original system considered by Unruh, acoustic flows [14], to more general considerations of dispersive media. The idea of dumb holes can indeed, as will be exemplified later in this chapter and all along the present Thesis, be generalised to any medium where a flow can be made to have a gradient such that its velocity eventually exceeds that of waves in the medium.

2.1.3.3 Analytical Description of Dumb Holes

Having developed an intuitive understanding of the idea behind the analogy of dispersive flows with the kinematics of spacetime in the vicinity of a black hole horizon, it is possible to establish analytically the connection between these two systems by deriving the wave equation of acoustics in flowing fluids. In order to proceed analytically, however, we must constrain our considerations to a nondispersive medium, as in the initial paper by Unruh [14]. As we will see shortly, this nonetheless captures the main ideas of the dumb hole proposal. Fluid dynamics are ruled by the equation of continuity[2]

$$\partial_t \rho + \nabla \cdot (\rho v) = 0 \tag{2.40}$$

and Euler's equation

$$\rho \frac{dv}{dt} \equiv \rho \left(\partial_t v + (v \cdot \nabla) v \right) = F, \tag{2.41}$$

with ρ the fluid density, v the flow velocity vector field, t the time as measured in the laboratory frame, and F the sum of all forces exerted on the fluid [15]. Following on the above discussion and after the work of Unruh [14], the flow is taken to have zero viscosity with the only forces being those due to the pressure p as well as Newtonian gravity and arbitrary external driving forces. The latter two are accounted for by the potential ϕ, yielding

$$F = -\nabla p - \rho \nabla \phi. \tag{2.42}$$

wherefrom the Euler equation 2.41 is rewritten as

$$\partial_t v = -\frac{1}{\rho} \nabla p - \nabla \phi - (v \cdot \nabla) v. \tag{2.43}$$

The enthalpy h of the barotropic fluid can be defined as a function of the pressure as $h(p) = \int_0^p \frac{dp'}{\rho(p')}$ so that $\nabla h = \frac{1}{\rho} \nabla p$. Furthermore, for a vorticity free flow, a veloc-

---

[2]Note that in this section, the partial derivative with respect to a variable is denoted by $\partial_t \equiv \frac{\partial}{\partial t}$ (only when greek indices μ, ν are written do we use the relativistic-covariant formulation).

ity potential ψ can be introduced such that $v = -\nabla\psi$ [16] so as to reduce Euler's equation to

$$\nabla \cdot \left(-\partial_t\psi + h + \frac{1}{2}(\nabla\psi)^2 + \phi \right) = 0. \tag{2.44}$$

In order to linearise the equations of motion (2.40) and (2.44) around a background, the exact motion $(\rho,\ p,\ \psi)$ is separated into an average background motion $(\rho_0,\ p_0,\ \psi_0)$ and low amplitude acoustic disturbances $(\rho_1,\ p_1,\ \psi_1)$ [15]. The continuity equations (2.40) for the background and acoustic disturbances are

$$\partial_t\rho_0 + \nabla \cdot (\rho_0 v_0) = 0, \tag{2.45}$$

$$\partial_t\rho_1 + \nabla \cdot (\rho_0 v_1 + \rho_1 v_0) = 0. \tag{2.46}$$

Similarly, the barotropic condition can be used in linearising the Euler equation (2.44), resulting in the pair

$$\partial_t\psi_0 + h_0 + \frac{1}{2}(\nabla\psi_0)^2 + \phi = 0. \tag{2.47}$$

$$p_1 = \rho_0 \left(\partial_t\psi_1 + v_0 \cdot \nabla\psi_1 \right) \tag{2.48}$$

Additionally, the barotropic assumption $\rho_1 = \frac{\partial\rho}{\partial p} p_1$ gives

$$\rho_1 = \frac{\partial\rho}{\partial p}\rho_0 \left(\partial_t\psi_1 + v_0 \cdot \nabla\psi_1 \right). \tag{2.49}$$

Substituting the latter consequence of the linearised Euler equation finally reveals the wave equation that describes the propagation of the linearised scalar potential ψ_1, that is that of acoustic disturbances:

$$-\partial_t \left(\frac{\partial\rho}{\partial p}\rho_0 \left(\partial_t\psi_1 + v_0 \cdot \nabla\psi_1 \right) \right) + \nabla \cdot \left(\rho_0\nabla\psi_1 - \frac{\partial\rho}{\partial p}\rho_0 v_0 \left(\partial_t\psi_1 + v_0 \cdot \nabla\psi_1 \right) \right) = 0. \tag{2.50}$$

Unruh identified the local speed of sound as $\frac{1}{c^2} \equiv \frac{\partial\rho}{\partial p}$ and realised that Eq. (2.50) could be cast into the equation of motion for a massless scalar field in a spherically symmetric geometry with static (inverse) metric [14] (note this is in $3 + 1$D)

$$g^{\mu\nu} = \frac{\rho_0}{c^2} \begin{pmatrix} -1 & \vdots & -v_0^j \\ \cdots & \cdot & \cdots \\ -v_0^i & \vdots & c^2\delta^{ij} - v_0^i v_0^j \end{pmatrix} \tag{2.51}$$

which compares with the Painlevé–Gullstrand metric (2.38) near the horizon of a black hole. There is a sonic horizon for the acoustic disturbance ψ_1 where the local speed of sound equals that of the background fluid ψ_0. This is a remarkable result: the velocity flow of a fluid can be analogous to a curved background for sound waves provided there is a flow-velocity gradient to supersonic flow. All the results of this Thesis build upon this finding.[3]

We will now turn away from fluids and sounds and discuss the translation of Unruh's idea of analogues to another setup: light in an optical fibre.

2.2 Developments in Optical Event Horizon Realisation

Light in optical fibres can be made to interact with itself to create a flowing medium. This section of the Thesis presents the tools of fibre optics that we rely upon to create those light-fast fluids and the event horizons they feature.

2.2.1 Fiber Optics

Light is an attractive experimental setup: it exhibits quantum properties at any temperature and is well understood and studied. We will here review the essential physics of light in optical fibres, swiftly progressing from linear optics to the lowest-order nonlinear response. We will conclude with the introduction of the essential constituent to the realisation of optical event horizons—the physics of the interaction of a soliton with a weak probe wave.

2.2.1.1 Nonliner Optics and Pulse Propagation

When a weak light wave interacts with a single body, the charged particle of a molecule or an atom, the wavelength of the output wave is identical to that of the input wave. Through the interaction with the molecule, the wave can suffer from attenuation, dispersion, deflection or be delayed, but the characteristic frequency of

---

[3] I will here seize the opportunity to remark that I find phenomenal that such an analogy is not only a powerful metaphor that aids the understanding of manifestations of black hole physics but an actual mathematical equivalence of the kinematics between the astrophysical and fluid systems. In order to regain perspective on this finding, I will insist on what I already wrote in the introduction to this dissertation: the fact that the propagation of sound waves in fluids is described by a curved Lorentzian manifold does not solely enable laboratory-based study of certain features of gravity. Rather, this incidentally (and probably even more importantly) also forces us to seriously rethink what we know of the connections between gravity, General Relativity, Einstein's equations, and the difference between kinematics and dynamics in this realm.

the waves will remain unaffected. The regime of weak light-matter interaction is ruled by the linear optics approximation.

A monochromatic electromagnetic radiation propagates in an optical fibre with phase velocity $v_p = \frac{c}{n}$, where c is the speed of light in vacuum and n the refractive index of the fibre material (of its core). The ultra short pulses that are usually sent in optical fibres are however not monochromatic but broadband: they are made of a wave at the carrier frequency and then numerous other waves at other frequencies around the carrier and are set by the mode spacing determined by the laser cavity that travel under an envelope. This envelope propagates with velocity $v_g(\omega) = \frac{c}{n_g(\omega)}$, the group velocity. We call $n_g(\omega)$ the (frequency-dependent) group index and define it as

$$n_g(\omega) = n(\omega) + \omega \frac{\partial n}{\partial \omega}, \qquad (2.52)$$

it explicitly depends upon the frequency of the wave, and therefore so does v_g. The group velocity of the wave packet is different from (and can be lower than) the phase velocity of the various frequency components it is made of (for the phase velocity is also frequency dependent). The phenomenon of frequency-dependence of the group and phase velocities—they differ in a dispersive medium—is what is called *dispersion*. It stems from material and waveguide dispersion, that is the frequency-dependence of, respectively, the refractive index and the size of the mode in the fibre. The latter depends mainly on the dimension of the core of the fibre. As a consequence of group-velocity dispersion, different spectral components of the pulse, because they will experience different refractive indices, will travel at different speeds. This results in a temporal broadening of the propagating pulse: the intensity of the pulse will be dispersed. The overall spectrum of the light packet is however not affected, for the pulse only spreads in time. Dispersion thus appears to tie in with the linear approximation. Three regimes of dispersion exist

- Normal dispersion where the high frequency (short wavelength) components travel faster than low frequency ones,
- Anomalous dispersion where the low frequency (long wavelength) components travel faster than the high frequency ones,
- The point of zero dispersion.

Dispersion leads to a walk-off between the different spectral components of the ultrashort light pulse: by (2.52), different frequency components of the pulse will travel at different group velocity in the medium. Thus the pulse broadens in the time domain and its peak intensity decreases. This eventually limits the efficiency of nonlinear effects: because two spectral components will propagate at different speed, the total electric field leads to lower nonlinearity strength.

In the situation where the light field impinging upon the above-mentioned molecule has a high intensity, the output wave will have a different frequency. In fact, if the intensity of the light is high enough, a pulse will induce effects on its own phase and amplitude through interaction with the molecule, resulting (by the virtue of Fourier Transform) in a change in its frequency. This regime of interaction is called nonlinear optics, because it stems from the nonlinear scattering of light and the nonlinear nature of the refractive index of the material. In order to understand why this is, let us explore a simple model of light and matter interaction. We model the medium in which light propagates, any dielectric such as silica—of which fibres are made of—for example, as a collection of charged particles. These basically are composed of light electrons bound to heavy ions. Maxwell's equations then allow us to interpret the propagation of light in this medium as propagating disturbances of the electric and magnetic fields. The polarisation of these fields depends upon the response of the bound charges within the medium to the electric field. In simple words, as an oscillating electric field travels in the material, the electric charges oscillate, inducing an electric dipole and radiating light (this is not instantaneous) at the driving frequency. In linear optics, when the light is of weak intensity, the polarisation of the medium is a linear response to terms of the first power of the electric field only. Hereafter we will only be interested in intense light fields that induce an anharmonic motion of the bound electrons of the material through propagation. This results in the polarisation of the medium becoming nonlinear and light being radiated at harmonic frequencies of the fundamental driving and wave mixing. For the purpose of this thesis, we will only consider phenomena belonging to the perturbative regime intensity interval (intensities of the order of 10^{11} to 5×10^{13} W.cm^{-2}). In this regime we consider the charged particles—the electrons—to be bound to the atom nucleus.

Pulse propagation in optical fibres obeys Maxwell's equations [17–19]

$$\nabla \times \mathbf{E} = -\frac{\partial \mathbf{B}}{\partial t},$$

$$\nabla \times \mathbf{H} = \mathbf{J} + \frac{\partial \mathbf{D}}{\partial t}, \qquad (2.53)$$

$$\nabla \mathbf{D} = \rho_f,$$

$$\nabla \mathbf{B} = 0.$$

where $\mathbf{E}$ and $\mathbf{H}$ are the electric and magnetic fields, respectively, $\mathbf{J}$ is the free current density, ρ_f is the free charge density, and $\mathbf{B}$ and $\mathbf{D}$ are related to $\mathbf{E}$ and $\mathbf{H}$ by the constitutive relations

$$\mathbf{D} = \epsilon_0 \mathbf{E} + \mathbf{P},$$

$$\mathbf{B} = \mu_0 \mathbf{H} + \mathbf{M}. \qquad (2.54)$$

$\mathbf{P}$ and $\mathbf{M}$ are the polarisation and magnetisation, respectively. The latter is zero in optical fibres—these are non-magnetic—and so are $\mathbf{J}$ and ρ_f—fibres are non conducting and electrically neutral. Substituting the constitutive relations into Maxwell's equations yields, via elimination of $\mathbf{D}$ and $\mathbf{B}$,

$$\nabla \times \mathbf{E} = -\mu_0 \frac{\partial \mathbf{H}}{\partial t},$$

$$\nabla \times \mathbf{H} = \epsilon_0 \frac{\partial \mathbf{E}}{\partial t} + \frac{\partial \mathbf{P}}{\partial t},$$

$$\nabla \mathbf{E} = -\frac{1}{\epsilon_0} \nabla \mathbf{P}, \tag{2.55}$$

$$\nabla \mathbf{H} = 0.$$

Taking the curl of the first equation allows for eliminating the magnetic field. We have successfully combined Maxwell's equations for light in a fibre to obtain the following propagation equation which involves the light field and the polarisation that is generated by the propagation of this light field:

$$\nabla \times (\nabla \times \mathbf{E}) + \mu_0 \epsilon_0 \frac{\partial^2 \mathbf{E}}{\partial t^2} = -\mu_0 \frac{\partial^2 \mathbf{P}}{\partial t^2} \tag{2.56}$$

The product $\mu_0 \epsilon_0 = 1/c^2$ with c being the speed of light in vacuum. As mentioned previously, the polarisation features both a linear and a nonlinear component:

$$\mathbf{P}(r, t) = \mathbf{P}_L(r, t) + \mathbf{P}_{NL}(r, t). \tag{2.57}$$

The linear (nonlinear) component of the polarisation is accounted for through the first (second) term of Eq. 2.57. The linear part of the polarisation describes the dispersion of the medium while the nonlinear part describes the nonlinear effects, through the first and second two terms of the following equation, respectively:

$$\mathbf{P}(r, t) = \epsilon_0 \chi^{(1)} \mathbf{E}(r, t) + \epsilon_0 \chi^{(2)} \mathbf{E}^2(r, t) + \epsilon_0 \chi^{(3)} \mathbf{E}^3(r, t) \tag{2.58}$$

where we have Taylor-expanded the electric field of the nonlinear terms and $\chi^{(k)}$, $(k) = (1, 2, \ldots)$, is the kth order of susceptibility of the medium. The dominant contribution to the polarisation induced by electric dipoles is provided by the linear susceptibility $\chi^{(1)}$ (which is related to the refractive index by $n^2(\omega) = 1 + \chi^{(1)}(\omega)$). A medium such as silica has inversion symmetry at the molecular level (silica is said to be centro-symmetric) and thus zero second order susceptibility $\chi^{(2)} = 0$, making the third order susceptibility responsible for the lowest-order nonlinear effects in optical fibres.

2.2.1.2 Optical Pulse

Let us now decompose the electric field associated with short laser pulses into the product of a modal distribution (spatial distribution of the electric field inside the waveguide) times a temporal envelope and the carrier frequency.

$$E(r, z, t) = F(r)A(z, t)e^{i\omega_0 t} \begin{cases} F(r), & \text{modal distribution} \\ A(z, t), & \text{temporal envelope} \\ \omega_0 = \frac{2\pi c}{\lambda_0}, & \text{carrier frequency} \end{cases} \qquad (2.59)$$

In an experiment, we actually measure the power, and not the amplitude, of the electromagnetic field $P(z, t) = |A(z, t)|^2$, that is the modulus squared of the envelope). Typically, a short laser pulse has a peak power ranging from some hundred watts to hundreds of kilowatts and can be as short as a few femto seconds. Depending on the peak power and duration of the pulse, the nonlinear effects observed in a fibre can be very different. As for the modal distribution, it is typically accurate enough to consider the propagation of the fundamental mode only and to assume its shape to be Gaussian. Because the transverse energy distribution of the fundamental mode does not vary through propagation, we generally study the change in shape of the temporal envelope and disregard the effects of diffraction.

Translating the above considerations into the pulse propagation equation (2.56), one arrives at the Generalized Nonlinear Schrödinger Equation[4] (GNLSE) that accounts for the combined effects of Loss (first term), Dispersion (second term) and all the nonlinear effects we will elaborate upon in the next paragraph (terms on the right-most side of the equation) [20]:

$$\frac{\partial A}{\partial z} + \frac{\alpha}{2}A - \sum \frac{i^{k+1}}{k!}\beta_k \frac{\partial^k A}{\partial T^k} =$$
$$i\gamma\left(1 + i\tau_{shock}\frac{\partial}{\partial T}\right)\left(A(z,t)\int_{-\infty}^{+\infty} R(T')\left|A\left(z, T - T'\right)\right|^2 dT'\right). \qquad (2.60)$$

Equation (2.60) governs the evolution of the field amplitude $A(z, t)$ (in units of $W^{-1/2}$) expressed in a frame of reference moving at the group velocity $v_g = 1/\beta_1$ of the pulse envelope such that $T = t - z/v_g$. The nonlinear coefficient γ (in units of $W^{-1}.m^{-1}$) describes the strength of nonlinear effects. It is related to the effective mode area[5] of the electric field in the fibre A_{eff} by $\gamma = \omega_0 n_2/cA_{eff}$—the strength of the nonlinear effects depends on the intensity and confinement of the electric field in the fibre (n_2 is the nonlinear refractive index, see discussion below). The frequency dependence of the mode area, and thus of γ, is described by the term τ_{shock}. The function $R(T) = 1 - f_r + f_r h_R(T)$ describes the nonlinear response and it includes both the instantaneous contribution (the Kerr effect) and delayed response ($h_R(T)$, Raman scattering). The coefficient f_r represents the Raman fractional contribution to the overall nonlinear response [20].

---

[4] This equation remains valid down to the single cycle regime, when the temporal envelope contains only one single oscillation of the electromagnetic field.

[5] The effective mode area is $A_{eff} = \frac{\left(\int |E|^2 dA\right)^2}{\int |E|^4 dA}$, where E is the electric field amplitude. The integration is done over the whole plane of the cross-section of the fibre. For a Gaussian beam with radius w, the effective area is πw^2 [20].

The terms on the left hand side of the equation account for the linear propagation effects via attenuation (α) and the dispersion of the propagation constant, while the terms on the right hand side describe the nonlinear effects. Let us detail the latter effects: the parentheses on the right of the GNLSE (2.60) describes the temporal envelope Self-Steepening, γ conveys the Self-Phase Modulation (SPM) and Four-Wave Mixing (FWM) effects whilst the integral describes the Stimulated Raman Scattering (SRS):

- *SPM*: Self-Phase Modulation is the effect where a polarized field of light modulates its own phase. This arises from the frequency dependence of the refractive index and results in phase shifts in the electric field. The frequency of the pulse being time-dependent, one observes the appearance of chirp (change in the instantaneous frequency across the pulse) on a ghost image, see Fig. 2.2. The temporal profile of the spectrum (its time-domain envelope) is not affected by this effect, only the spectrum broadens. One could say that this is the opposite effect to dispersion: the SPM-induced chirp is similar to that caused by normal dispersion, such that longer wavelengths propagate faster within the pulse and are located on the leading edge.
- *FWM*: Another consequence of the intensity-dependence of the refractive index is the so-called Four Wave Mixing process, which consists in a nonlinear mixing between two optical signals at different frequency and the resulting generation of signals at the frequency difference and sum of the frequencies. Figure 2.3 clearly shows that the frequency of the four waves involved in the process add up and that energy is thus conserved. An efficient process where a consequent part of the energy of the two initial waves is transferred to the newly generated signals requires the phase of the waves to be matched: all the waves have to be in phase so that

$$\beta(\omega_1) + \beta(\omega_2) = \beta(\omega_3) + \beta(\omega_4) \tag{2.61}$$

- *SRS*: The third nonlinear effect encrypted in the GNLSE is Stimulated Raman Scattering, which stems from the interaction between light and the vibrational modes of the molecules. This interaction yields a Raman gain to be produced for a wave with a shorter frequency than the high intensity pump in the case of Stokes scattering.[6] The Raman gain profile is very broadband and depends on the material, Fig. 2.4 depicts it for silica.

The GNLSE can be made to account for the effects of noise, for the frequency-dependence of the mode area, and for the wave polarisation. This equation is extremely powerful and widely used in nonlinear fibre optics [20]. The pulse propagation dynamics strongly depends upon the pump wavelength relative to the zero dispersion wavelength (ZDW): at low wavelength, in the normal dispersion region, SPM will be observed while FWM actually takes place at the ZDW for example. The regime of interest for us is that of anomalous dispersion which allows, through modulation instability, for the generation of solitons in the fibre.

---

[6] Anti-Stokes scattering—frequency up-conversion—is also possible, although Stokes scattering is more frequency [20].

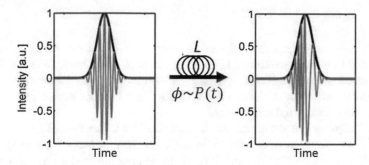

Fig. 2.2 Self phase modulation: under the Kerr effect, an intense pulse modifies the refractive index of the fibre material as it propagates. This results in the pulse experiencing an additional phase shift beside the linear phase shift. In other words, light modulates its own phase: $\phi_{NL}(t, L) = \gamma P(t) L$, where L is the propagation distance of the pulse and γ the nonlinear coefficient of the fibre. The black line corresponds to the pulse envelope and the grey line the amplitude of the electric field. Figure adapted from [20]

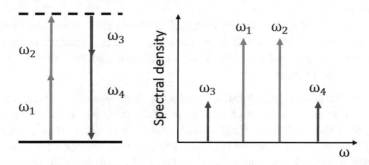

Fig. 2.3 Four wave mixing: FWM is the interaction of four waves with distinct frequencies via a third-order nonlinearity. It describes the annihilation and/or generation of four distinct photons

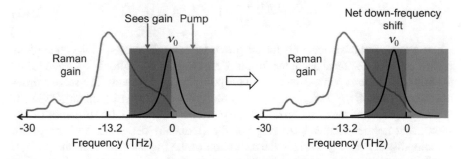

Fig. 2.4 Stimulated raman scattering: raman scattering corresponds to the energy transfer from photons to phonons by an inelastic collision. Through collision the photon energy is changed: it thus frequency shifts. The Raman gain spectrum for fused silica, at $\lambda = 1060$ nm, is plotted after [20]

2.2.1.3 Solitons

We will now investigate the different regimes of soliton propagation in order to understand in what context optical solitons are created in fibres. In that spirit, we will proceed to deconstruct the dynamics of soliton propagation step by step and show how the different nonlinear effects we introduced earlier work together to give rise to a massive spectral broadening.

Four "regimes" of dynamics can be identified as a function of the dispersion regime. The evolution of a short pulse (of high peak power, of the order of ten kilowatts) propagating in the anomalous dispersion regime can be divided into three stages:

- First the higher-order soliton is being compressed, which results in a spectral broadening,
- The pulse then splits into a range of small pulses, fundamental solitons. At the same time, the spectrum starts expanding under the effects of dispersive wave generation.
- Finally, the wavelength of the fundamental solitons shifts to longer wavelengths under the effect of Raman self-frequency shift. As a consequence of this wavelength shift, the peak shifts and the pulse is no longer symmetric or well described by a Gaussian envelope.

Now that the soliton propagation dynamics has been carefully deconstructed, this paragraph will present a thorough description of the three evolution stages, starting with the fundamental soliton. Fundamental solitons essentially are stable solutions of the nonlinear Schrödinger equation that appear when the chirp from self-phase modulation balances that of anomalous dispersion. This is described by the following simplified Nonlinear Schrödinger equation (NLSE):

$$i\frac{\partial A}{\partial z} + \frac{|\beta_2|}{2}\frac{\partial^2 A}{\partial T^2} + \gamma|A|^2 A = 0 \tag{2.62}$$

where the second term accounts for the group velocity dispersion in the anomalous regime and the last term accounts for the Kerr effect (the nonlinear dependence of the refractive index upon the intensity of the pumping mechanism). The second order coefficient $\beta_2 = \partial^2\beta/\partial\omega^2|_{\omega_0}$ is known as the group velocity dispersion (GVD). It governs the rate of temporal broadening experienced by the pulse. Equation (2.62) shows that neither Raman scattering nor the frequency dependence of the group velocity dispersion play a role in the formation of the fundamental soliton.

Mathematically, it is possible to show that, in the anomalous dispersion regime ($\beta_2 < 0$), a certain class of solutions may fulfil Eq. (2.62): these are called fundamental solitons. Requirements for this are twofold, fundamental solitons must have [20]:

- a hyper-secant shape, this is mathematically described by

$$A\,(z = 0, T) = \sqrt{P_0}sech\,(T/T0)\,, \quad A\,(z, T) = \sqrt{P_0}sech\,(T/T0)\,e^{iz\gamma P_0/2},$$
(2.63)

T_0 is the duration of the pulse and P_0 is the peak power.
- a soliton number N—which determines the maximum energy for which the interplay between dispersion and nonlinearity allows for a stable solution—that must be unity:

$$N = \sqrt{L_d/L_{NL}} = \sqrt{\gamma P_0 T_0^2/|\beta_2|} = 1$$
(2.64)

where, $L_d = T_0^2|\beta_2|$ and $L_{NL} = 1/(\gamma P_0)$ the dispersion and nonlinear lengths, respectively.

Fundamental solitons are not the only solution to the NLSE, it actually allows for higher order solitons to exist. These have an integer soliton number higher than fundamental soliton (e.g. $2, 3, 4, \ldots$). They correspond to the interference of fundamental solitons, with different amplitudes and phase, during propagation. In this case, the soliton is periodic upon propagation:

$$A\,(z + L_d, T) = A\,(z, T)\,, \quad A\,(z = 0, T) = \sqrt{P_0}sech\,(T/T0)\,.$$
(2.65)

Here, L_d is the dispersion length, and β_2 is the group-velocity dispersion parameter of the medium at the carrier frequency. Upon propagation, a second order soliton, for example, goes through a cycle of expansion and compression in the time domain (and reciprocally compression and expansion in the spectral domain). Nothing binds fundamental solitons together to form higher-order solitons: they only have the same velocity and thus interfere constructively. If the degeneracy of the velocities of the different constituents of a higher-oder soliton is disturbed, that is if these constituents start propagating at different velocities in the fibre, the interference between them will not be constructive any more. A soliton of order $N > 1$ will therefore break into N fundamental solitons.

2.2.1.4 Pulse and Probe Interaction

The theory developed for this Thesis accounts for the interaction of a soliton with a weak probe wave, or with waves in the quantum vacuum state (not populated with photons). We will now proceed to describe such an interaction. For the sake of this Thesis, it is enough to consider cross-phase modulation only. Consider a fundamental soliton and a continuous-wave probe of intensity significantly lower than that of the pulse forming the soliton. Their well-separated central frequencies are denoted ω_s and ω_p, respectively. In practice, the nonlinear interaction between the two fields is unidirectional with the soliton acting on the probe and the back-reaction being negligible because of the low intensity of the second field. The nonlinear polarisation, which is in the same direction as the probe field E_p, reads [21, 22]

$$P_{NL} = \frac{r}{2}\epsilon_0 \chi^{(3)} |E_s|^2 E_p, \tag{2.66}$$

where r is 1 if the fields are orthogonally polarized and 3 if the fields are polarized along the same direction. With such a polarisation, the full wave equation in real space is [21, 23][7]

$$c^2 \left(\partial_z^2 + \beta^2(i\partial_t)\right) E_p = \partial_t^2(\chi E_p) = 0. \tag{2.67}$$

Identifying

$$\chi = \frac{r}{2}\chi^{(3)} |E_s|^2 \tag{2.68}$$

as the nonlinear susceptibility induced by the pulse and writing the propagation constant $\beta = \omega n(\omega)/c$ out, we find that this wave equation can be written as

$$c^2 \partial_z^2 E_p - \partial_t^2 n^2 (i\partial_t) E_p - \partial_t^2 (\chi E_p) = 0, \tag{2.69}$$

where from it becomes clear that the susceptibility χ induced by the pulse on the probe is a local change of refractive index. We identified this effect earlier as the Kerr effect (see Sect. 2.2.1).

Let us recall the dynamics of propagation of a wave of low intensity in a bare fibre. Although the refractive index of the fibre experienced by the wave depends on its frequency, it is constant along propagation. A weak probe of constant frequency will therefore propagate with constant group and phase velocity. The wavepacket propagation is a bit more complex because of dispersion—the difference of the above mentioned velocities for different frequencies—but the refractive index profile of the medium is not modified by the probe. And its frequency is thus unchanged. This is in contrast with the influence of the soliton on the probe: under the Kerr effect it will modify the susceptibility and the probe will experience a local change of refractive index,

$$n_{eff}^2 = n^2 + \chi. \tag{2.70}$$

This transient increase in refractive index will have implications on the probe field: its group and phase velocities will change. And under dispersion, it will frequency shift. This transient frequency shift, usually almost negligible, can be significant in the case where the probe and soliton have very small relative speed.

This change in velocity and frequency of a probe wave interacting with a pulse—soliton—in an optical fibre are the fundamental ingredients of optical event horizons. Contrarily to 'dumb holes' the inhomogeneity of waves in the medium is not induced by a change in the fluid flow velocity but by a local change in the index of the

---

[7]Note that in this section, the partial derivative with respect to a variable is denoted by $\partial_t \equiv \frac{\partial}{\partial t}$ (only when greek indices μ, ν are written do we use the relativistic-covariant formulation).

medium—hence a change in the waves velocity with respect to the medium. In the next section, we shall elaborate upon the first experiment that demonstrated a classical effect of analogue event horizons in an optical setup: frequency shifting at the group velocity horizon.

2.2.2 Fibre-Optical Analogue of the Event Horizon

In this section we will examine further the scheme of an intense pulse propagating in an optical fibre, and is effect on weak probe waves. We will see how, when the group velocity of the probe and pulse are matched, the probe experiences a group velocity reversal in the co-moving frame, in addition to a frequency shift. This, as was discussed in the River Model of the black hole and its implications (see Sect. 2.1.3.1), is the behaviour of waves at analogue event horizons. We will show how this analogy for optical waves was first demonstrated.

2.2.2.1   At the Speed of Light

The influence of the pulse on the probe wave is best described in a frame of reference in which the pulse is stationary. For the purpose of the present discussion, we will adopt the coordinate transformation of [23]:

$$\zeta = \frac{z}{u}, \ \tau = t - \frac{z}{u}, \tag{2.71}$$

with u the group velocity of the pulse—and hence the speed of the frame. Note the peculiar implications of this coordinate transformation: the propagation distance has been normalized with respect to the velocity, which yields the propagation time. To the exchange of space and time, this transformation is similar to a Galilean transformation. Although ζ and τ do not correspond to space and time coordinates in the moving frame (the transformation is not a Lorentz transformation), they do define a non-inertial coordinate system in which we can solve the wave equation. As the frame moves at speed u, it is clear that the pulse remains centred at $\tau = 0$. We assume a pulse whose energy distribution does not vary along propagation—that is independent of ζ. This implies that the nonlinear susceptibility is simply $\chi = \chi(\tau)$. Expressing the optical wave equation in terms of the vector potential A_p, related to the electric and magnetic fields via $E_p = -\partial_t A_p$ and $B_p = \partial_z A_p$, and substituting the partial derivatives of the coordinates in the co-moving frame ($\partial_z = \frac{1}{u}(\partial_\zeta - \partial_\tau)$, $\partial_t = \partial_\tau$) into the wave equation (2.69), we find the wave equation for weak probe waves in the presence of an intense pulse

$$(\partial_\zeta - \partial_\tau)^2 A_p + u^2 \beta^2 (i\partial_\tau) A_p - \frac{u^2}{c^2} \partial_\tau^2 (\chi \partial_\tau A_p) = 0. \tag{2.72}$$

Thanks to the exchange of the space and time coordinates, the simple form of the operator $\beta^2(i\partial_t)$ is maintained. In the historical method, the authors derived this wave equation from a Lagrangian.

The wave equation will be solved by decomposing a solution into its plane wave components. The latter behave according to the dispersion relation of the fibre in the co-moving frame [21, 23]

$$\omega' \approx \omega \left(1 \mp \frac{u}{c} \left(n(\omega) + \frac{\chi}{2n(\omega)} \right) \right),$$ (2.73)

with ω' the frequency as measured in the comoving frame. We see that the nonlinearity affects the form of the dispersion profile but not the velocity—u is constant in this model. The dispersion relation is a second order polynomial, it has two branches corresponding to probe waves propagating forward or backward in the moving frame—co- or counter-propagating, respectively, with respect to the pulse in the laboratory frame.

2.2.2.2 Fibre-Optical Analogy

Before solving the wave equation (2.72), we explain how a soliton in a fibre acts as a pair of analogue horizons to probes waves. The propagation of a short and intense pulse in an optical fibre simulates a flow-velocity profile that moves with the pulse. Each pulse locally increases the effective refractive index of the fibre because of the Kerr effect. The speed of light waves in the fibre being determined by the refractive index, this pulse behaves like a perturbation in the flow velocity of the fibre. A wave propagating in the same direction as the pulse attempts to move against this flow, through interaction with the pulse its velocity decreases. To waves in the fibre, the pulse separates two regions of flow velocity: the outside of the pulse, where it moves at subluminal speed, from under the pulse, where it moves superluminal. This is in complete analogy with a moving fluid. Clearly, both edges of the pulse act as an event horizon: at the trailing edge of the pulse the flow velocity decreases in the direction of the pulse, this is a white hole horizon. The opposite occurs at the leading edge of the pulse, this is the time reverse of a white hole horizon, a black hole horizon.

Frequency shifting at the white hole event horizon being easier to explain, we will first focus on the fate of a wavepacket incident upon the trailing edge of the pulse. This wavepacket will experience an increase in refractive index and thus be slowed down as it approaches the pulse. The front of the wavepacket will interact with the pulse before its back does, and will thus be slowed down earlier—resulting in a compression of its wavefront, an increase in frequency. In the absence of dispersion, this would continue indefinitely and the wavepacket would be squashed at the horizon with ever-increasing frequency. This reminds us of the effect of a white hole on a wave emitted from its horizon (the time-reverse of the black hole infinite redshift): an infinite blueshift, and of the Trans-Planckian problem with Hawking radiation.

Fortunately, this shifting will be limited by dispersion. Indeed, because it depends on dispersion, as its frequency increases the speed of the wavepacket reduces, until it becomes slower than the pulse. In the co-moving frame, the wavepacket then appears to turn around and to be dragged away from the horizon. An analogue white hole event horizon thus reflects and blueshifts the incoming wavepacket. Owing to the time-reversal symmetry of the black- and white-hole horizon effects on waves, we can now simply state that a black hole would redshift and reflect an incoming wavepacket.

As we will show later (see Sect. 4.3.1) any transient increase in the refractive index forms an event horizon for optical frequency waves. The value of the frequency shift of the incoming wavepacket depends on the magnitude of this increase, and can be easily determined: the process of frequency shifting must conserve the co-moving energy $\hbar\omega'$ (because the Lagrangian describing the interaction is time-invariant in the co-moving frame). We write the dispersion in a form that explicitly illustrates the dependence of ω' on δn ($\delta n \approx \chi/2n$):

$$\omega' = \omega \left(1 - \frac{u}{c}\left(n(\omega) + \delta n(\omega, \tau)\right)\right). \tag{2.74}$$

Note that the τ dependent change induced by the pulse has been included in agreement with Refs. [21, 23]. According to this equation, $\omega' = $ constant is a condition that determines a family of contours in the $\omega - \delta n$ plane (see an example in Fig. 2.5). During the interaction, ω' lies on one of these contours (determined by ω the laboratory frequency of the wavepacket and δn). Reflection in the co-moving frame, which is accompanied by frequency shifting in the laboratory frame, is possible when the initial frequency of the wave in the laboratory frame is close to ω_m, the group-velocity-matching frequency (the frequency which has the same group velocity as the pulse). At ω_m the group velocity in the moving frame vanishes. Around ω_m, the $\omega' = $ constant contours form parabolas centred at ω_m. An incident wavepacket will thus see its laboratory frequency be modified by the pulse, as it travels along a parabolic contour and lands at the shifted frequency—that has moving frame group velocity opposite to that of the incident frequency. Waves outside the event horizon frequency window[8] will be able to pass through the pulse.

This geometrical optics description does unfortunately not describe *entirely* the interaction of the probe and pulse. Indeed, real waves are spatially extended and do not have well-defined values of both ω and τ. A full wave treatment of the optical field (see Refs. [21, 23]) reveals that waves are only partially reflected at the pulse. Some of the wave energy can be transmitted via the tunnelling effect [24]. The amount of reflection decreases as the incident wave frequency moves away from ω_m. We shall elaborate further on this in Sects. 3.2 and 5.3.

---

[8]cf. Sect. 3.2.3.

Fig. 2.5 Frequency shifting contours: each contour corresponds to a single co-moving frequency, ω_0', in the δn plane. These contours form parabolas centred at the group-velocity-matching frequency, ω_m. Figure and caption from [21]

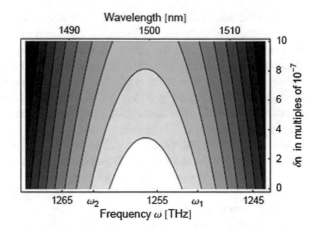

2.2.2.3 Frequency Shift at the Horizon

After the waveguide-based proposal [25], a collaboration between the Quantum Optics groups of Leonhardt and König, at the University of St Andrews, was the first to propose a model for analogue horizons in optical fibres realised by means of a fundamental soliton propagating in a fibre. They experimentally demonstrated the frequency shifting of waves at a white hole horizon in a seminal paper published in 2008 [23]. A full quantum treatment of the field accompanying this publication established that pairs of photons are emitted from the vacuum at the horizons formed by the edges of a soliton in the fibre.

If the probe is replaced in the fibre by a set of sufficiently weakly excited modes (even in the state of quantum vacuum), these will experience the cross Kerr effect of the pulse [23]. These modes constitute a quantum field of light, and light is a real electromagnetic wave so, according to Fourier analysis, their oscillations at positive angular frequency ω will be accompanied by the complex conjugate amplitude at $-\omega$. These positive- and negative-frequency modes of the field have a positive- and negative-norm, respectively, in the field theory [23, 26, 27]. At the event horizon in the fibre, these modes will mix, thus creating observable light quanta (a more detailed analysis supporting this statement will be the topic of Chap. 4).

The St Andrews team did not observe the spontaneous emission of light from the vacuum at a horizon but showed that the expected temperature of emission would be of the order of 10^3 Kelvin, many orders of magnitude higher than any other condensed matter analogue system promises. The scheme they designed benefits from the fact that all the aspects of the physics of analogue event horizons come together to facilitate the observation of Hawking Radiation.

2.2.2.4 An Optical Wave with Negative Frequency?

Before we move on to presenting the novel theory of spontaneous emission of light quanta at a moving horizon, let us dwell upon the idea of negative frequency waves, as we did in the introduction for the one-dimensional string. A collaboration inspired by an original idea of Friedrich König indeed reported having observed the transfer of energy from a soliton to a dispersive wave of negative frequency [28].

All light oscillate with both positive and negative frequencies: the field A is related to its frequency spectrum $\tilde{A}$ by Fourier transformation, recall Eq. (1.1) from the introduction—

$$A = \int_{-\infty}^{+\infty} \tilde{A}(\omega) \exp^{-i\omega t} d\omega, \qquad (2.75)$$

where the integral extends from negative to positive frequencies. Since A is a real electric field of light, $A^* = A$, hence $\tilde{A}(-\omega) = \tilde{A}^*(\omega)$. Accordingly, the negative part of the spectrum entirely depends upon the positive part, which makes negative frequencies seem redundant for waves. However, it is possible to perform an experiment where the positive frequency part couples to the negative frequency part, thus displaying the full complex nature of the electromagnetic field [29].

In quantum physics, the positive- and negative-frequency components of the field are assigned a positive and a negative *pseudo*-norm, respectively, by (1.12). In the field expansion (see Sects. 3.1.2 and 3.2.2 for details), the positive norm mode of the field is attached to its annihilation operator, and the negative norm mode to its creation operator. In a process where positive norm modes are coupled to negative norm modes (and vice versa), the creation and annihilation operators of the field will mix, which results in the spontaneous creation of photons from the vacuum.

In the experiment [28], a temporal soliton is propagated along an optical fibre. In the presence of higher order dispersion, the fundamental soliton becomes unstable and can, in this case, couple to dispersive waves that have the same momentum as the soliton. This resonance effect is known as Čerenkov radiation [30–32]. The generated wave is normally of positive frequency (norm), but the momentum conservation also allows for a negative frequency (norm) solution.

The momentum conservation can be expressed in terms of the Doppler shifted frequency in the moving frame of the soliton, $\omega' = \omega - uk$, where u is the velocity of the soliton and $k = n\omega/c$. The resulting condition is $\omega' = \omega'_{IN}$, where ω' (ω'_{IN}) is the frequency of the generated light (input soliton). Momentum conservation in the laboratory frame corresponds to energy (frequency) conservation in the moving frame. For the sake of the present argument, we consider a simplified two branches (positive and negative optical laboratory frequency) dispersion relation of light in the comoving frame, as displayed in Fig. 2.6. The figure shows that, under the Doppler effect, some parts of the laboratory positive and negative frequency branches have positive co-moving frequency. Thus, there are two further laboratory frequency ω that share the same ω'_{IN} with the input soliton. One is of positive laboratory-frame frequency and is the above mentioned positive frequency resonant radiation (RR)— or Čerenkov radiation—and the second is of distinctly negative laboratory-frame

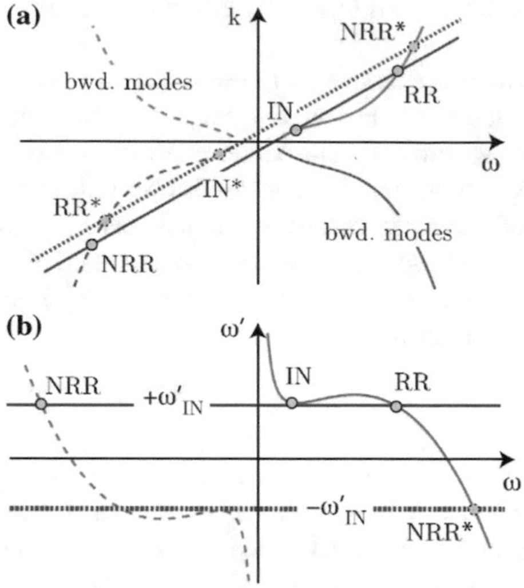

Fig. 2.6 Typical dispersion relation $k = k(\omega)$, e.g., for fused silica glass with second- and third-order dispersion, **a** in the laboratory reference frame and **b** in the reference frame comoving at the soliton velocity. Dashed curves indicate the (laboratory frame) negative frequency branches of the dispersion relation. Figure and caption from [28]

Fig. 2.7 Experimental results for negative RR generation in a photonic-crystal fiber. **a–b** Measured spectra in the visible and UV regions for three different input energies: 246 pJ (dotted line), 324 pJ (dashed line), and 366 pJ (solid line). **c** Full fiber dispersion relation: positions of the predicted RR and negative RR spectral peaks are indicated. The inset is a 25× enlargement of the curve around the RR wavelength. Figure and caption from [28]

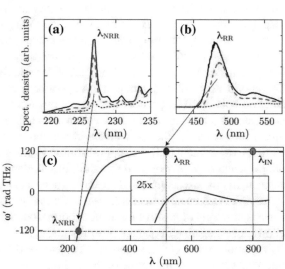

frequency. The authors of [28] call this wave the negative resonant radiation (NRR). Note that there is no positive solution at minus the negative frequency, except for the complex conjugate fields, enforcing that the field is real-valued (see Appendix A).

For the experiment, a 7-fs nJ-energy pulse was coupled into a few mm-long fibre. The pulse compresses in the fibre with a very wide spectrum such that it excites both the RR and NRR modes. In the laboratory frame, the conjugate field mode to the NRR lies in the UV, around 230 nm depending on the fibre used, very far from the IR-centred pulse. Thus the energy transfer is less efficient between the pulse and the NRR than it is between the pulse and the RR. Nevertheless, a clear signal can observed at the expected wavelength (see Fig. 2.7). It is strongest when $\omega'_{IN} = \omega'_{soliton}$, that is when energy is transferred from the central laboratory frequency component of the soliton. Furthermore, it was established that the generation of NRR light depends on the pulse compression and fibre parameters: the excitation of the NRR mode critically depends on the spectral support in the ultraviolet [33].

The generation of NRR, via its quantum field origins and also because it ought to fulfil conditions akin to those of light scattering on an optical event horizon (see Sect. 2.2.2), shed light on the physics of astrophysical particle creation in optical analogues. It also promises to be an excellent tool to test the theory developed in Chap. 4.

2.3 Equations of the Optical Spacetime

Before delving into the details of the quantum field theory used in this Thesis, let us briefly examine the arguments that support the analogy between light in media and black hole physics. We will here develop a simple, and yet completely relativistic, theory of light propagation in dispersive media. This is a novel result of this Thesis, and goes beyond works in the literature that have solely considered the equations for light in nondispersive media (see, for example, [23, 34–36]) or regimes of dispersion without group velocity dispersion [37] to draw the analogy with the sonic metric found by Unruh [14] (2.51). Thus, we develop a different formalism from those presented previously to explain the formation of a soliton in the fibre or to establish the existence of horizons in dispersionless media: indeed, the present argument does not depend upon the details of the mechanism that underlies the existence of the horizon condition. Rather, we seek a phenomenological understanding drawn from the mathematics of General Relativity, the peculiarities of the wave equation we call upon are only a means to this end.

2.3.1 Action in an Optical Medium as an Analogue Metric

Recall the Painlevé–Gullstrand metric (2.38), in $1 + 1D$ it is

$$ds^2 = -\left(c^2 - \beta^2\right) d\tau^2 + d\zeta^2 + 2\beta d\tau d\zeta. \tag{2.76}$$

We wish to establish under which conditions the action of light in an optical medium can be analogous to that of a massless scalar field in (2.76). To this end, we now develop a completely relativistic theory of light propagation in dispersive, inhomogeneous media.

So far in the present chapter, we have always considered that the electromagnetic field depended only on the longitudinal and time coordinates, z and t. This is motivated by the fact that in the experiments that are relevant to our theory, light (for example the pulse and the probe of Sect. 2.2.2) propagates in the z direction, and the variations of the electromagnetic field in the transverse directions effectively are negligible. For example, in an optical fibre, light propagation is based on total internal reflection. If we consider the set of transverse modes along y, which is discrete, we can see that, for small y, the energy of the modes with $k_y > 0$ is so large that these transverse modes can be neglected ($\lambda >> \Delta y$). Thus, the electromagnetic field $A\,(z, t)$ does not depend on y (the transverse coordinate). That is, the propagation of the electromagnetic field can be described by an effective action in a $1 + 1$ dimensional space.

In a regime of linear dispersion, for frequencies much smaller than the resonance frequency of a medium ($|\omega| \ll \Omega$), the (low-energy) action of the electromagnetic field $A(z, t)$ on the medium

$$\mathcal{L} = \frac{1}{2} \left(\left(1 + \frac{4\pi\kappa}{\Omega^2} \right) (\partial_t A)^2 + c^2 (\partial_z A)^2 \right) \tag{2.77}$$

leads to the dispersion relation[9]

$$c^2 k^2 = \omega^2 \left(1 + \frac{4\pi\kappa}{\Omega^2} \right). \tag{2.78}$$

κ can be understood as being related to the coupling strength of light in the medium. Both Ω and κ depend on z and t in the laboratory frame.

We here face a problem: the field equation resulting from (2.77) is conformally invariant, we cannot introduce an analogue effective geometry in $1 + 1$ dimensions. To circumvent this, we use the 'silent' (extra-)dimension y[10] to write an analogue metric [25] of line element[11]

$$ds^2 = -c^2 dt^2 + \left(1 + \frac{4\pi\kappa}{\Omega^2} \right) (dy^2 + dz^2). \tag{2.79}$$

Equations (2.77) and (2.79) are the keys to the study of the Lorentzian manifold that describes the optical spacetime. We may now investigate how the curvature enters this manifold, or, in other words, under which conditions the propagation of light in dispersive, inhomogeneous media can be analogous to motion on a curved gravitational background, and when the curvature is such that there is an event horizon.

---

[9]We shall derive and explain the full version of this equation in Chap. 3. Note that Eq. (2.78) is an approximation valid when the medium features only one resonance.

[10]There is no direct relation between the metric and the wave equation for light in media.

[11]Note that in this section, the partial derivative with respect to a variable is denoted by $\partial_t \equiv \frac{\partial}{\partial t}$—we do not use the relativistic-covariant formulation.

2.3.2 A Black Hole Horizon for Light

We can find under which conditions the modes of oscillation of the electromagnetic field in the medium will experience an event horizon by pushing the analysis of our General Relativity toy model Eq. (2.79) further. For this purpose, we rewrite the metric tensor of Eq. (2.79)[12] as

$$g = \begin{pmatrix} -c^2 & 0 & 0 \\ 0 & 1 + \frac{4\pi\kappa}{\Omega^2} & 0 \\ 0 & 0 & 1 + \frac{4\pi\kappa}{\Omega^2} \end{pmatrix} \tag{2.80}$$

and express this metric in a stationary form by transforming it via

$$dt = \gamma \left(d\tau + \frac{u}{c^2} d\zeta \right), \; dz = \gamma \left(d\zeta + u d\tau \right), \tag{2.81}$$

with $\gamma = 1/\sqrt{1 - u^2}$ the Lorentz factor, to

$$ds^2 = -\left(c^2 - u^2 \left(1 + \frac{4\pi\kappa}{\Omega^2} \right) \right) \gamma^2 d\tau^2 + \left(\left(1 + \frac{4\pi\kappa}{\Omega^2} \right) - \frac{u^2}{c^2} \right) \gamma^2 d\zeta^2$$

$$+ 2u\gamma^2 \frac{4\pi\kappa}{\Omega^2} d\zeta d\tau + \left(1 + \frac{4\pi\kappa}{\Omega^2} \right) dy^2 \tag{2.82}$$

The transformation from (2.79) to (2.82) consisted in a Lorentz boost to an inertial frame moving at velocity u with respect to that in which the action was initially considered—the latter will be referred to as the observer's frame in the remaining of the section. In the stationary form (2.82), the metric is similar to (2.76), the Schwarzschild solution to Einstein's vacuum equation as expressed by Painlevé [11] and Gullstrand [12]. As we saw in Sect. 2.1.2.1, this metric has a horizon when the Newtonian β velocity of space equals the speed of light c, for which $g_{00} = 0$.

In regions of the dispersion relation where ω is approximately linear in k, in which (2.78) is valid, the g_{00} component is

$$g_{00} \propto c^2 - u^2 \left(1 + \frac{4\pi\kappa}{\Omega^2} \right) = 1 - \frac{u^2}{v_p^2}, \tag{2.83}$$

where we have identified $v_p = c \left(1 + 4\pi\kappa/\Omega^2 \right)^{-1/2}$—the phase velocity of waves in the observer's frame. Thus, there is a black hole when $v_p = u$, in total analogy with the black hole metric (2.76)!

---

[12]The line element and metric tensor are related by Eq. (2.24).

Clearly, this can only be achieved if Ω and κ, the material properties, are not constant in spacetime. By assumption, the material properties are independent of time: $\partial_\tau \Omega = \partial_\tau \kappa = 0$. So one would necessarily resort to a *moving spatial disturbance* in the medium—a moving change in the refractive index—to create the conditions such that $v_p = u$. The preferred frame from which to boost from the observer's frame is then obviously that of the moving disturbance (that propagates through the medium at constant speed u). For the sake of the argument, say we are able to increase the refractive index of the medium over a finite spatial region, thus creating a Refractive Index Front (RIF) that propagates at speed u (in the positive z direction in the observer's frame). Furthermore, for simplicity, let us assume that only the resonant frequency Ω is affected (decreased) by this increase in the refractive index (that is, κ remains constant throughout the medium, even under the RIF, and Ω depends on ζ). Then, by studying the dispersion relation (2.78) one finds that the phase velocity of light decreases as the refractive index increases. In other words, light is slower under the RIF. The latter can then act as a black hole event horizon for modes of light that would have the adequate frequency in the observer's reference frame.

In a dispersive medium, the group and phase velocity of light modes are different. Thus there will also be a condition similar to $g_{00} = 0$ for the group velocity: $v_g = u$. This is not well described by the simple toy model (2.78) and the study of this condition is therefore postponed until the next chapter of this dissertation.

2.3.3   Conclusion and Discussion

In this section of the Thesis, we have presented the theory supporting the science of optical spacetime realisation. From first principles, Maxwell's equations for electromagnetic waves, we have established a wave equation for light in optical fibre. We have explored under which conditions a few-cycles and intense light pulse coupled in an fibre could create a soliton, which, via the Kerr effect, modifies the refractive index of the fibre, thus creating a flow velocity profile. It was shown how a weak probe co-propagating with the soliton will experience a transient change in refractive index and how this bears classical features of gravity—black and white holes. In particular, we elaborated upon the generation of waves with negative frequencies in the laboratory at the event horizon, a promising observation that inspired the work presented herein. In the next chapter we will use the tools of quantum field theory to explain what the, so far mysterious, Hawking radiation phenomenon is, and how light can be spontaneously emitted from the vacuum at the event horizon. We will then use a quantum theory based on a more involved version of the light-matter interaction model (2.78) to calculate the properties of emission from the quantum vacuum at the group velocity horizon created by a moving disturbance in the refractive index of a highly dispersive dielectric.

References

1. C.W. Misner, K.S. Thorne, J.A. Wheeler, *Gravitation*. (W.H. Freeman, San Francisco, 1973)
2. A. Einstein. Die feldgleichungen der gravitation, in *Sitzungsberichte der Preussischen Akademie der Wissenschaften zu Berlin* (1915), pp. 844–847
3. A. Einstein, Die grundlage der allgemeinen relativitatstheorie. Ann. der Phys. **354**(7), 769–822 (1916)
4. Y. Choquet-Bruhat, *Introduction to General Relativity, Black Holes, and Cosmology* (Oxford University Press, Oxford, first edition edition, 2015)
5. K. Schwarzschild, Uber das gravitationsfeld eines massenpunktes nach der einsteinschen theorie. Sitz. der K. Preuss. Akad. der Wiss. **7**, 189–196 (1916)
6. Nature a comparison of whitehead's and einstein's formulae. **113**(2832), 192–192 (1924)
7. G. Lemaitre, L'Univers en expansion. Ann. de la Soc. Sci. de Brux. **A53**(51) (1933)
8. D. Finkelstein, Past-future asymmetry of the gravitational field of a point particle. Phys. Rev. **110**(4), 965–967 (1958)
9. R. Penrose, Black holes and gravitational theory. Nature **236**(5347), 377–380 (1972)
10. E.F. Taylor, J.A. Wheeler, *Exploring Black Holes: Introduction to General Relativity*. (Addison Wesley Longman, San Francisco, 2000)
11. P. Painleve, La mecanique classique et la theorie de la relativite. C. R. Acad. Sci (Paris) **173**, 670–680 (1921)
12. A. Gullstrand, Allgemeine lesung des statischen einkerperproblems in der einsteinschen gravitationstheorie. Arkiv. Mat. Astron. Fys. **16**(8), 1–15 (1922)
13. J.S. Hamilton, P. Lisle, The river model of black holes (2006)
14. W.G. Unruh, Experimental black-hole evaporation? Phys. Rev. Lett. **46**(21), 1351–1353 (1981)
15. L.D. Landau, E.M. Lifšic, *Fluid Mechanics*. Number v. 6 in Course of theoretical physics. 2nd edn. (Pergamon Press, Oxford, England; New York, 2nd english edn., rev edition, 1987)
16. M. Visser, Acoustic black holes: horizons, ergospheres and hawking radiation. Class. Quantum Gravity **15**(6), 1767 (1998)
17. J.C. Maxwell, On physical lines of force. Phil. Mag. 11, 11611–175; 281–291; 338–348 (1861)
18. J.C. Maxwell, On physical lines of force. Phil. Mag. **12**(12–24), 85–95 (1862)
19. J.C. Maxwell, *A Treatise on Electricity and Magnetism*. (Clarendon Press edition, 1873)
20. G. Genty, M. Narhi, C. Amiot, M.J. Jacquet, Supercontinuum generation in optical fibers, in *Proceedings of the International School of Physics Enrico Fermi* (2016), pp. 233–261
21. S. Robertson, Hawking radiation in dispersive media. Ph.D. thesis, University of St Andrews, St Andrews, 2011
22. M. Jacquet, Quantum Vacuum emission at the event horizon. M.Sc. thesis, University of St Andrews, St Andrews, 2013
23. T.G. Philbin, C. Kuklewicz, S. Robertson, S. Hill, F. König, U. Leonhardt, Fiber-optical analog of the event horizon. Science **319**(5868), 1367–1370 (2008)
24. A. Choudhary, F. König, Efficient frequency shifting of dispersive waves at solitons. Opt. Express **20**(5), 5538 (2012)
25. R. Schützhold, W.G. Unruh, Hawking radiation in an electromagnetic waveguide? Phys. Rev. Lett. **95**, 031301 (2005)
26. R. Brout, S. Massar, R. Parentani, P. Spindel, A primer for black hole quantum physics. Phys. Rep. **260**(6), 329–446 (1995)
27. N.D. Birrell, P.C.W. Davies, *Quantum Fields in Curved Space*, repr edn. Cambridge monographs on mathematical physics. (Cambridge Univ. Press, Cambridge, 1994)
28. E. Rubino, J. McLenaghan, S.C. Kehr, F. Belgiorno, D. Townsend, S. Rohr, C.E. Kuklewicz, U. Leonhardt, F. König, D. Faccio, Negative-frequency resonant radiation. Phys. Rev. Lett. **108**(25) (2012)
29. M. Conforti, A. Marini, T.X. Tran, D. Faccio, F. Biancalana, Interaction between optical fields and their conjugates in nonlinear media. Opt. Express **21**(25), 31239 (2013)
30. N. Akhmediev, M. Karlsson, Cherenkov radiation emitted by solitons in optical fibers. Phys. Rev. A **51**, 2602–2607 (1995)

31. P.K.A. Wai, C.R. Menyuk, Y.C. Lee, H.H. Chen, Nonlinear pulse propagation in the neighbor-hood of the zero-dispersion wavelength of monomode optical fibers. Opt. Lett. **11**(7), 464–466 (1986)
32. D.V. Skryabin, A.V. Yulin, Theory of generation of new frequencies by mixing of solitons and dispersive waves in optical fibers. Phys. Rev. E **72**, 016619 (2005)
33. J.S. McLenaghan, Negative frequency waves in optics: control and investigation of their gen-eration and evolution. Ph.D. thesis, University of St Andrews, St Andrews, 2014
34. U. Leonhardt, P. Piwnicki, Optics of nonuniformly moving media. Phys. Rev. A **60**(6), 4301–4312 (1999)
35. U. Leonhardt, T.G. Philbin, The case for artificial black holes. Philos. Trans. R. Soc. A: Math. Phys. Eng. Sci. **366**(1877), 2851–2857 (2008)
36. D. Faccio, S. Cacciatori, V. Gorini, V.G. Sala, A. Averchi, A. Lotti, M. Kolesik, J.V. Moloney, Analogue gravity and ultrashort laser pulse filamentation. EPL (Europhys. Lett.) **89**(3), 34004 (2010)
37. M.F. Linder, R. Schützhold, W.G. Unruh, Derivation of hawking radiation in dispersive dielec-tric media. Phys. Rev. D **93**(10) (2016)

Chapter 3
Spontaneous Emission of Light Quanta from the Vacuum

3.1 Quantum Field Theory in Curved Spacetime

In the previous chapter of this dissertation we used the classical theory of Physics that rules the dynamics of the Universe on large scales—General Relativity—to study the behaviour of spacetime around spherical bodies. We introduced the idea of black holes, regions of spacetime bounded by their event horizon from which nothing can escape. In this section, we will try to tie General Relativity with Thermodynamics— broadly speaking, the theory that rules the organization of the Universe. For this purpose, we will follow the arguments which scientists of the early 1970s had to contend with, and see how they found that these theories can be united at the event horizon of black holes. This will eventually lead us to call upon Quantum Physics to explain how black holes can be in a state of thermal equilibrium—thus introducing the concept of spontaneous emission of light quanta from the vacuum.

The structure of this section is inspired by that of the series of seminars I gave to PhysSoc, the undergraduate society at the School of Physics and Astronomy in St Andrews, in the Autumn of 2016. The material presented here builds on the content of these seminars, although the treatment will be much more mathematical and more room will be dedicated to considerations drawn from General Relativity and, ultimately the quantum theory of fields in curved spacetime. There exists a large body of work that treats this material in different ways, see for example Carter's 1973 review [1], Davies' 1978 review [2], Birrell and Davies' 1982 book [3], or Jacobson's 1996 lecture notes [4].

3.1.1 Gravity and Thermodynamics: The Failure of Classical Physics

The difficulty in classically describing the interaction between the conceptually dissimilar aspects of fundamental physics accounted for by Thermodynamics and

© Springer International Publishing AG, part of Springer Nature 2018
M. J. Jacquet, *Negative Frequency at the Horizon*, Springer Theses,
https://doi.org/10.1007/978-3-319-91071-0_3

Gravity arises from the apparent absence of true equilibrium in astrophysics. This is exemplified by the observation that a star is not made hotter by adding matter to it but by removing matter from it—contrarily to laboratory thermodynamic systems— they radiate and get hotter[1] (like all self-graviting systems [5]). The self-gravitation effects of stars is only compensated for by their internal pressure (that arises from their internal kinetic or zero-point quantum pressure) and their temperature (that arises from thermonuclear fusion). In that regard, stars are a metastable state of matter in the history of the Universe: would a solar-mass star loose all its heat energy, it would undergo a dramatic shrinkage to a fraction of its initial size and, after a period of oscillations, explode in a nova resulting in the formation of a higher temperature cloud of gas. Heavier stars would undergo gravitational collapse and become black holes. In any case, a star is a mere, and timely, interlude of matter organisation between a distended cloud of gas and imploded matter.

We will begin by defining precisely the meaning of the event horizon and thus pose the problem with black hole entropy. This will lead us to the analogy between the laws of Thermodynamics and those ruling the size of black holes and to the formulation of four laws for black hole mechanics. We will rely on the concept of information [6] to identify the surface gravity of the hole with its temperature. In terms of classical physics, this procedure leads to a paradox: given that the interior of the black hole is hidden and inaccessible to us (outside observers) [7], it would imply that it has a high entropy. Indeed, we cannot tell what has formed the hole—it being characterised completely by its mass, angular momentum and electric charge, it can be the result of the gravitational collapse of an infinite number of initial configurations [8]. The information about the internal microstates that the initial star was composed of is wiped out by the collapse to leave only information about the macro state (characterised by the three global parameters mass, angular momentum, and electric charge) to be measured by an external observer.[2] Thus, on physical grounds, it appears that the bigger the hole, the more information it would have wiped out when collapsing. This seems to indicate that the size of the hole provides a measure of its entropy. We will see how this paradox can be 'solved' by calling on quantum physics to give a meaning to the temperature attached to this entropy.

3.1.1.1 Black Holes and Their Event Horizon

In the previous chapter, we studied the structure of spacetime around a spherical body in the framework of General Relativity. Stars having exhausted their nuclear fuel will shrink under their own gravity (because their inner pressure can no longer compensate their weight). Chandrasekhar calculated that any object heavier than approximately 1.39 times the mass of the Sun at the onset of shrinkage [9] could not become a white dwarf—having no low temperature equilibrium it would become a

---

[1]The nuclear mass loss associated with fusion results in energy release that heats up the star.

[2]In that sense, the black hole represents the state of maximal entropy, that is the equilibrium end state of gravitational collapse.

neutron star and/or undergo complete gravitational collapse. We have seen that no signal could travel outwards from singularities which occur in gravitational collapses, the latter being hidden behind the event horizon—not visible to an outside observer.

Based on the mathematical framework of General Relativity, we saw how a black hole on a spacelike surface could be referred to as a connected component of the region of the surface bounded by the event horizon. Here, we will begin by summarising the properties needed to study black holes. We will then discuss the region outside a collapsed body in terms of these properties and, by studying possible trajectories for photons (null geodesics), establish where there is an event horizon. We will follow the steps of Hawking [8] and postulate the existence of stationary black holes and prove that real solutions to the field equations tend towards these. From there, we will progress to establishing that the event horizon does indeed have a spherical topology.

A star having exhausted its nuclear fuel will undergo gravitational collapse. If the collapse is exactly spherically symmetric, the metric is that of the Schwarzschild solution outside the star (2.15). It has the following properties:

(i) The surface of the star will shrink inside the Schwarzschild radius $r_S = \frac{2GM}{c^2}$. When this happens, the spacelike 2-surface at r_S will be such that both the future directed families of null geodesics orthonormal to it are converging—it will be a closed marginally trapped surface. The star will be in such a strong gravitational field that even outgoing light from it will be dragged inwards.

(ii) There is a spacetime singularity.

(iii) The singularity is not visible to observers at $r > r_S$. This implies that one can predict the future in the exterior region from the initial conditions with respect to the time parameter (Cauchy data) on a spacelike surface.

Work by Penrose [10], Hawking and Ellis [7], and Gibbons [11] in the early 1970s showed that these three properties hold. Elaborating upon these properties, we will proceed to show that the surface area of the event horizon cannot decrease with time.

In order to discuss the region outside a collapsed object, one needs a precise notion of infinity in an asymptotically flat spacetime. This was provided by Penrose's concept of a weakly asymptotically simple space [12]. The spacetime manifold $\mathcal{M}$ of such a space can be embedded in a larger, Lorentzian, manifold $\tilde{\mathcal{M}}$ with a metric conformal to that of $\mathcal{M}$—$\tilde{g}_{ab} = \Omega^2 g_{ab}$. The function Ω is smooth and zero, with non-vanishing gradient, on the boundary of $\mathcal{M}$ in $\tilde{\mathcal{M}}$. This boundary consists of two null hypersurfaces $\mathcal{F}^+$ and $\mathcal{F}^-$ which each have topology $S^2 \times \mathcal{R}^1$: these represent the future and past null infinity respectively.[3]

Let us define the partial Cauchy surface $\mathcal{S}$, a spacelike surface without edge which does not intersect any non-spacelike curve more than once, and $\mathcal{D}^+(\mathcal{S})$, the set of all points p such that every—extended enough—past directed non-spacelike curve from p intersects $\mathcal{S}$. $\mathcal{D}^+$ is called the future Cauchy development of $\mathcal{S}$.

---

[3]The future of a set is the collection of all spacetime points that can be reached by future-going timelike or null curves from that set.

Proposition (iii) above states that it should be possible to predict events near $\mathcal{F}^+$. That is, a weakly asymptotic space is (future) asymptotically predictable if S is such that $\mathcal{F}^+$ lies in the closure in $\mathcal{M}$ of $\mathcal{D}^+$. In other words, a space is asymptotically predictable if there are no singularities in $J^+(S)$, the future of S, which are naked, i.e., not surrounded by an event horizon of finite radius. This is mathematically expressed by saying that there is no singularities which lie in the past of future null infinity $J^-(S)$.[4]

If we consider an asymptotically predictable space in which there are no singularities to the past of S, and suppose there is a closed trapped surface[5] T in $\mathcal{D}^+(S)$, then there will be a non-spacelike geodesic in $J^+(S)$ which is future incomplete and cannot be seen from the end point $\mathcal{F}^+$. That is, there will be a singularity to the future of T; as T is a trapped surface, the null geodesics orthogonal to T are converging.

The past of future null infinity of S, $J^-(S)$, thus does not contain T: in topological terms, its boundary $j^-(\mathcal{F}^+)$ is the event horizon for $\mathcal{F}^+$ [8]. $j^-(\mathcal{F}^+)$ is generated by null geodesic segments which have no future end-point—it is the boundary of the region from which particles or photons can escape to infinity. Let us call $S(t)$, $(t > 0)$ a family of partial Cauchy surfaces[6] in $\mathcal{D}^+(S)$. For sufficiently large t, $S(t)$ will intersect the event horizon: $\mathcal{B}(t) = S(t) - J(\mathcal{F}^+)$ will be non-empty. A black hole is then a region of $S(t)$ from which there is no escape to $\mathcal{F}^+$, a connected component of $\mathcal{B}(t)$ [8].[7] The study of spacetimes which possess an event horizon reduces to the study of the event horizon.

If the collapse was strongly asymptotically predictable, one would also expect the solution of the field equations outside the event horizon to become stationary at late times. Armed with this intuition, Hawking postulated stationary solutions to the field equations outside the event horizon and proved that real solutions do indeed tend towards these [8]—thus placing certain limits on the possible behaviour of black holes. An immediate consequence of the definition of the event horizon as the boundary of a past is that through each point of the horizon surface there passes a maximally extended future-directed geodesic which remains always in the horizon—it never reaches $\mathcal{F}^+$. These null geodesics are called the generators of the horizon $j^-(\mathcal{F}^+)$. The convergence of these generators can never be positive (Hawking provided a proof by contradiction of this statement in [8]). This bears huge implication in terms of the possible behaviour of black holes: since the null geodesic segments that generate the event horizon have negative convergence and have no future end point, the surface area of the boundary of the black hole cannot decrease with time. Additionally, after Carter, Hawking showed that the event horizon

---

[4]The past of future null infinity of S, $j^-(\mathcal{F}^+)$, physically represents the set of all events from which an observer could escape to the asymptotic region.

[5]This is a closed, spacelike, 2-surface whose ingoing and outgoing null normal geodesics are both converging. For example, a sphere at constant r and v in Eddington-Finkelstein coordinates is a trapped surface if it lies inside the horizon.

[6]A partial Cauchy surface is a hypersurface which is intersected by any causal curve at most once.

[7]A spacetime in which certain observers can never escape to the asymptotic region, i.e., for which the past of future null infinity is not the entire spacetime, is a spacetime that has an event horizon. It is said to possess a black hole.

of a stationary black hole is a sphere—it has the topology $\mathcal{S}^2$ (even if two black holes merge into a bigger black hole, if this resulting black hole is stationary, topologically, its event horizon will be a sphere).

3.1.1.2 Laws of Black Hole Mechanics

As they were studying the interaction between two black holes as defined in the previous section, Bardeen and Hawking [13] derived the expressions for the mass of a stationary axisymmetric solution of the Einstein's equations for both a black hole surrounded by matter and for the difference in mass between two neighbouring such solutions. After their results and treatment, we will see how the area of the event horizon and the surface gravity, two quantities that appear in their result, are analogous to the thermodynamics concepts of entropy and temperature respectively. The argument will culminate in the formulation of four laws of black hole mechanics corresponding to the four laws of thermodynamics. This shall eventually allow us to progress to the paradox of thermal equilibrium of black holes.

Already in 1972, Hawking had uncovered the analogy between thermodynamics and black holes [8]: according to the theory laid out in the previous section, the surface area of the event horizon of a black hole cannot decrease with time, i.e. $\delta A \geq 0$.[8] This is analogous to the *second law* of Thermodynamics, which states that the entropy of a system always increases with time.[9]

Let us digress for a moment and consider the behaviour of a particle outside the event horizon of a black hole. If the particle rigidly corotates with the black hole, it will have some angular velocity, a 4-velocity vector as well as an acceleration 4-vector. The (redshifted) amplitude of the acceleration tends to some constant when the particle is infinitesimally close to the event horizon.[10] This constant can be thought of as the *surface gravity* κ of the black hole [13].

We can now call on Bardeen's and Hawking's finding [13] that any two neighbouring stationary axisymmetric solutions containing a perfect fluid with circular flow and a central black hole, whose event horizon has a surface area A, and of angular momentum J_H, are related by the differential mass formula:

$$\delta M = \frac{\kappa}{8\pi}\delta A + \Omega_H \delta J_H + \int \Omega \delta dJ + \int \bar{\mu}\delta dN + \int \bar{\theta}\delta dS, \qquad (3.1)$$

where δdJ is the change in the angular momentum of the fluid crossing an infinitesimal surface element, and δdN and δdS are the change in the number of particles in—and in entropy of—the fluid crossing the same surface element. $\bar{\mu}$ and $\bar{\theta}$ are

---

[8] δA is the change in surface area of the event horizon of the black hole.

[9] Note that it was Hawking who discovered that black hole horizons must grow *if* there is only positive energy that falls in, and Bekenstein who later established the link between this observation and entropy.

[10] The acceleration of the particle arbitrarily close to the horizon goes to infinity, but from afar this is multiplied by the redshit factor, which also tends to infinity in this case, yielding a finite constant.

the "red-shifted" chemical potential and temperature of the fluid. Ω_H is the angular velocity of a particle outside the event horizon which corotates rigidly with the black hole. Let us compare this equation with that of a microscopic non-reversible change in internal energy in terms of microscopic changes in entropy, and volume for a closed system in thermal equilibrium—the fundamental thermodynamic relation [14]:

$$dU = TdS - PdV + \sum_i \mu_i dn_i, \tag{3.2}$$

where the μ_i are the chemical potentials corresponding to particles of type i, and the usual (reversible and of constant chemical composition) thermodynamic relation has been generalised to account for potential change in the composition, i.e., the amounts n_i of the chemical components in the system. P and V are the pressure and volume, respectively, of the closed system of internal energy U. From the first term of (3.1), one can see that the quantity $\frac{\kappa}{8\pi}$ is analogous to the absolute temperature T in the same way that A is analogous to entropy S in (3.2). This is the *first law* of black hole mechanics.

Ascribing an effective temperature to the black hole did not shock Bardeen and Hawking: because time dilation factor tends to zero at the horizon (see Eq. 2.36), the redshifted temperature $\bar{\theta}$ of any matter orbiting the hole must tend to zero as the horizon is approached.

At the time, however, they opposed the above analogy with Thermodynamics temperature and entropy by the following argument: a black hole cannot be in equilibrium with black body radiation at any non zero temperature. Indeed, no radiation can be emitted from the black hole, whereas some radiation will always cross the horizon into the hole. Furthermore, they note that if one followed this analogy, any addition of entropy to a black hole would cause some increase in the area of the event horizon (which is classically constant).

Nonetheless, continuing the analogy between surface gravity and temperature, one can formulate the remaining two laws of black hole mechanics [8]. The *zeroth law* states that the surface gravity is constant over the event horizon, and the *third law* stresses that it is impossible to reduce this surface gravity to absolute zero by any procedure consisting of a finite sequence of operations.

3.1.1.3 The Paradox of Black Hole Heat

In the previous section, we established an analogy between the surface gravity of a black hole and the concept of temperature in thermodynamics. But how can a black hole have a temperature: it cannot emit anything, it cannot emit heat. And can thus not be in thermal equilibrium with incoming radiation. This is a *paradox*. Calling on Davies' idea that information can be equated with negative entropy [2], we will present the argument used by Bekenstein in 1973 [15] to establish the relationship between temperature and mass and arrive at the conclusion that classical physics fails to properly describe the thermodynamics of black holes.

Let us glance back at Eq. (3.1), and remark that it can be interpreted as an expression of mass-energy conservation (corresponding to the above *first law* of black hole mechanics). Now, after Smarr [16] and from (3.1), we express the total surface area of the horizon as a measure of size, thus writing:

$$A = 4\pi \left(2M^2 - e^2 + 2M^2 \left(1 - \frac{e^2}{M^2} - \frac{J^2}{M^4} \right)^{1/2} \right) \tag{3.3}$$

in units of $G = c = 1$, and with $e^2 < M^2$ and $J^2 < M^4$ (e the electric charge, J the angular momentum and M the mass of the black hole). In his 1972 theorem [8] upon which we dwelt earlier, Hawking showed that the horizon area cannot decrease (even for black holes having an electric charge and angular momentum), thus opening the route to the study of black hole thermodynamics. From (3.3), and for a Schwarzschild black hole ($e = 0$, $J = 0$), we can write

$$A = 16\pi M^2 \tag{3.4}$$

and

$$\kappa \equiv \frac{\partial M}{\partial A} = \frac{1}{4M}. \tag{3.5}$$

wherefrom we can evaluate the entropy of the hole: recalling that, by (3.1), $8\pi \frac{dM}{\kappa} = dS$, it suffices to integrate[11] $dM = \kappa dA = TdS$ to arrive at the thermodynamic relation:

$$2\frac{1}{8\pi 4M} S = 2ST \tag{3.6}$$

$$M = 2ST \leftrightarrow energy = 2\, entropy \times temperature$$

where the factor 2 entered as a result of the quadratic dependency of A on M. If we rewrite the thermodynamic relation (3.6) in the form of (3.4), we can express A and κ as the product of two finite quantities $M = A\kappa/4\pi$. As the energy M is finite, a black hole with zero temperature would seem to have *infinite* entropy. This is puzzling.

To understand this puzzle, we resort to the relation between entropy and information [2]: a highly ordered system has a low entropy, the amount of information needed to describe it is very large (it has a high information content). The explanation for this seemingly counter-intuitive statement goes as follows: the information about the microstates that initially composed the star is destroyed by the gravitational field (i.e., the space-time structure) and it becomes inaccessible to an outside observer because of the event horizon. Before it had collapsed, the star had an ordered and structured state, that was characterised by information about all the microstates that composed the star. Upon collapsing, the system changes from this ordered, structured state to a few-parameter disordered state—after the collapse, the black hole is

---

[11]The first law of black hole mechanics states that $S \leftrightarrow A$.

only characterised by the three global parameters $(M \ J \ e)$—and thus less information
is needed to describe it. As the order of a system decreases, its entropy increases:
less information is thus required to describe its state. A system in thermodynamic
equilibrium—such as the black hole, which is the state of maximal entropy of the
collapse—thus appears to be in the state of maximum entropy and minimum information content, a small number of parameters is needed to describe it. As a result,
information seems to correspond to *negative entropy*. Such considerations of information are useful in understanding the nature of the event horizon, as defined in a
previous paragraph: the ongoing discussion leads to the conclusion that black holes
possess a large entropy because of all the information they have wiped out. On the
basis of classical physics, the configurations and number of particles that have produced the black hole is infinite [17]. If we assign one bit of information to each
degree of freedom of these particles, we see that the information content of the black
hole, and hence its entropy, should indeed be infinite.

This unbounded entropy can be considered as being connected with the instability
of matter against total collapse. Just like one would apply quantum theory to an atom
(thereby ascribing it a stable ground state that prevents the electron from spiralling
indefinitely close the nucleus), it is fortunately possible to take into account the quantum nature of the matter that formed a hole. Let us now lay some heuristic arguments
on information and entropy down to arrive at an expression for the temperature of
a black hole. The relation between energy and wavelength of a particle, $E = h/\lambda$,
states that particles that produce the hole must have a wavelength shorter than the size
of the hole for their energy to be located within it. For the radius of a Schwarzschild
black hole, this leads to $\lambda \simeq 2M$, and thus a minimum particle mass of the order of
h/M, and hence to a maximum number of particles that went into forming this hole
of about M^2/h. The entropy can then be estimated to be

$$S = \xi k_B \left(\frac{M^2}{h} \right) \tag{3.7}$$

where k_B is the Boltzmann constant and ξ is a scaling factor for the entropy of the
hole. It is a dimensionless constant of order unity whose exact magnitude will be
uncovered in the next section. From (3.7), it is clear that S diverges in the classical
limit of $h \to 0$ (but is otherwise finite!): one needs to find a full quantum theory of
black holes to set a bound on their entropy. Bekenstein [15] showed that the entropy
is proportional to the area of the event horizon by rewriting (3.7) in the form of (3.4):

$$S = \frac{\xi k_B}{16\pi h} A. \tag{3.8}$$

From (3.6), we finally find the temperature of a Schwarzschild black hole to be

$$T = \left(\frac{h}{2\xi k_B} \right) M^{-1}. \tag{3.9}$$

By (3.5), this is

$$T = \left(\frac{2h}{\xi k_B} \right) \kappa, \tag{3.10}$$

that is, T is directly proportional to the surface gravity of the hole. The latter equation implies that a black hole would have to emit energy at the same rate as it absorbed it: that a hole has a temperature means that it is in thermal equilibrium with a surrounding heat bath at the same temperature. In other words, for the black hole to have a temperature we must associate with it a thermal equilibrium radiation spectrum. Having stated this, the immediate question to pose is that of the origin of this radiation—what is the mechanism behind it and where it originates from. To answer this, we will proceed to a fully quantum treatment of fields in the vicinity of black holes, but we can first make a few comments about the localisation of the radiation upon emission.

The characteristic wavelength of this radiation will be $\lambda_0 \simeq h/k_b T \simeq 2M$. If we assume ξ to be approximately unity, we find that this wavelength is of the order of the radius of the black hole (as one would have expected from our choice of entropy). This shows that the notion of location of the origin of the radiation (e.g. inside or outside the hole) bears no meaning. That is, a temperature can be associated with the black hole via Eq. (3.10) without having to state that radiation flows *out of the black hole itself*. Nonetheless, we do have to consider the black hole as being hot—a source of heat radiation. Classical physics fails at explaining this, we will therefore resort to a full quantum treatment to unveil the mystery of black hole radiation and solve the paradox of black hole heat.

3.1.2 Black Hole Evaporation

In this section, we will elaborate further on the final result of the above paragraph: that, according to Eq. (3.10), a black hole has a temperature. We will use the framework of relativistic quantum field theory to incorporate the effects of h, the Planck constant, in the theory, thus morphing the heuristic thermodynamic analogy into true thermodynamics. The science of analogue event horizons being similar in treatment to the historical approach to quantum gravity: semiclassical considerations of quantum fields in a fixed (classical) background—in the present case a black hole geometry—this section will be used to introduce the basic concepts and tools needed to undertake this venture.

In what follows, we will see how the vacuum fluctuations of a field in such a background have an effect on the thermodynamics of black holes via Hawking radiation. We will begin by studying the history of a light mode in a gravitational collapse. We will then show how distortions in the background geometry lead to Hawking radiation being emitted by the hole and dwell upon the physical origin of this flux.

3.1.2.1   History of a Light Mode in a Gravitational Collapse

In 1971, Penrose established that it was possible to extract rotational energy from
a black hole with infalling particles [18]. Zel'dovich [19] and Misner [20] then
showed that similar process for waves existed—*super-radiance*. Zel'dovich [21],
Starobinsky [22] and Unruh [23] identified this process with stimulated emission
and asked whether a rotating black hole would spontaneously radiate. In his efforts
in favour of spontaneous emission from rotating black holes, Hawking first found
that a non-rotating black hole would emit a thermal spectrum of particles [24].

In order to derive this result, it is only necessary to consider the case of a massless,
scalar field in the Schwarzschild spacetime (this can be generalised to any quantum
field in a general black hole spacetime, see for example [3]). So as to avoid the issue
of boundary conditions on the past horizon, we do not consider the full Schwarzschild
spacetime: we imagine that the black hole was formed at some time in the past—on
$\mathcal{F}^-$. In what follows, we will find the relationship between waves that propagated
from past null infinity J^- through the collapsing body—before the horizon formed—
and emerged from it, thus undergoing a very large redshift to J^+. For the outgoing
waves F on J^+ to have finite frequency, the incoming waves f must have left J^- with
very high frequency [25]: we can thus rely on the geometrical optics approximation
to describe their propagation on the background geometry.

In this scheme, a $v = constant$ ingoing ray, of pure positive frequency on J^-,
propagates through the collapsing body and emerges as a $u = constant$ outgoing
ray, of positive frequency on J^+. We relate u and v, the retarded time coordinates for
the modes of the field, by $u = g(v)$, or $v = g^{-1}(v) \equiv G(u)$—with $v = t + r^*$ and
$u = t - r^*$ (see Sect. 2.1.2). The two sets of modes have asymptotic form

$$f_\omega \approx \begin{cases} e^{-i\omega v}, & \text{on } J^- \\ e^{-i\omega G(u)}, & \text{on } J^+ \end{cases}, \quad F_\omega \approx \begin{cases} e^{-i\omega u}, & \text{on } J^+ \\ e^{-i\omega g(v)}, & \text{on } J^-. \end{cases} \tag{3.11}$$

We will adapt the general method provided by Birrell and Davies in their book
[3] to derive Hawking's result for the explicit case of a symmetric ball of matter
imploding across its event horizon. This is a one-dimensional analogue of gravitational collapse that will allow us to investigate the physics of the Hawking emission
mechanism close to the black hole. The ball has a thin shell, and, in its exterior
region, is surrounded by empty space. Thus the unique solution of Einstein's equation (2.14) is the Schwarschild spacetime described by the metric (2.15). Following
on the treatment presented in this chapter, we express the line element of this metric
in the Eddington-Finkelstein form (2.18) via the tortoise coordinate transformation
(2.17) and write the retarded space coordinate

$$r^* = r + 2M \ln(\frac{r}{2M} - 1). \tag{3.12}$$

Inside of the ball, on the other side of the thin shell, the properties of spacetime
are irrelevant [3, 25]. For simplicity, and contrarily to Birrell and Davies, we will

consider the spacetime to be flat: this will allow us to trace the history of the ingoing modes as they propagate through the ball and convert into outgoing modes. Inside the ball, spacetime is thus described by the Minkowski metric of line element (2.24)

$$ds^2 = -dT^2 + dr^2,   \tag{3.13}$$

We define $V = T + r$ and $U = T - r$ as the null coordinates constant on the interior region ingoing and outgoing rays, respectively.

We want to establish the relation between the incoming and outgoing rays on the black hole, therefore we let $r = R(t)$ describe the history of the shell (that will contract inside of its Schwarzschild radius). In this one-dimensional hypersurface, the metric must be the same as seen from both sides of the shell—this leads to two conditions: the intrinsic geometry must match and the extrinsic curvatures of each side of this hypersurface must match. The latter allows for determining $R(t)$ in terms of the stress-energy in the shell, which is not needed presently. We thus focus on the first condition, which reads:

$$dT^2 - dR(t)^2 = \left(1 - \frac{2M}{R(t)}\right) dt^2 - \left(1 - \frac{2M}{R(t)}\right)^{-1} dR(t)^2$$

$$\Rightarrow 1 - \left(\frac{dR(t)}{dT}\right)^2 = \left(1 - \frac{2M}{R(t)}\right)\left(\frac{dt}{dT}\right)^2 - \left(1 - \frac{2M}{R(t)}\right)^{-1}\left(\frac{dR(t)}{dT}\right)^2.$$
$$\tag{3.14}$$

We now use these matching conditions to determine the relation between the values of the null coordinates from $\mathcal{F}^-$ through the shell, v and V, through the centre of the ball, V and U, and through the shell again to $\mathcal{F}^+$, U and u—see Fig. 3.1. In other words, we will find the form of the modes of the field in the remote future, after incoming waves have converged on the centre of the ball, have passed through it to become outgoing waves and propagated to $\mathcal{F}^+$. We denote the limiting value of v for rays which pass through the ball before it has shrunk to the critical compactness (at $r = r_s$) as v_0.

For an incoming null rays entering the ball at a radius finitely larger than $2M$, both $\left(1 - \frac{2M}{R(t)}\right)^{-1}$ and $\frac{dR(t)}{dT}$ are finite, and approximately constant. Thus the $\frac{dt}{dT}$

Fig. 3.1 History of a ray passing through a collapsing ball. An ingoing ray v enters the collapsing ball, passes through the origin, and exits as an outgoing ray u

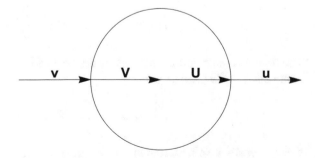

derivative is approximately constant, i.e. $t \propto T$. Likewise, in these condition r^* is linearly proportional to r. Hence the relation between v and V for $v = v_0$:

$$V(v) = av + b, \tag{3.15}$$

where a and b are constants. At the centre of the shell, $r = 0$; there, the above expressions for the null coordinates in the interior region become

$$U(V) = V. \tag{3.16}$$

Upon exiting the shell, at a time T_0 at which $R(t) = 2M$, $R(T) \approx 2M + A(T_0 - T)$ (where A is a constant). Inserting in the matching conditions (3.14) results in

$$\left(\frac{dt}{dT}\right)^2 \approx \left(\frac{R(t)}{2M} - 2M\right)^{-2} \left(\frac{dR(t)}{dT}\right)^2 \approx \frac{(2M)^2}{(T - T_0)^2}, \tag{3.17}$$

implying

$$t \sim -2M \ln\left(\frac{T_0 - T}{B}\right), \; T \to T_0 \tag{3.18}$$

(B is a constant). Likewise, as $T \to T_0$, r^* becomes

$$r^* \sim 2M \ln\left(\frac{r}{2M} - 1\right) \sim 2M \ln\left(\frac{A(T_0 - T)}{2M}\right), \tag{3.19}$$

hence

$$u = t - r^* \sim -4M \ln\left(\frac{T_0 - T}{2M\, B/A}\right). \tag{3.20}$$

Furthermore, in this limit, $U = T - r = T - R(T)$, so

$$U \sim (1 + A)\,T - 2M - AT_0. \tag{3.21}$$

From Eq. (3.20), and identifying that at $R = r = 2M$, $T = t$, one gets that $T = v_0 - r^*$. Similarly, $T_0 = v - r^*$, and thus

$$u = g(v) = -4M \ln(\frac{v_0 - v}{2M\, B/A}). \tag{3.22}$$

This is the same result as that obtained by Hawking with his general ray tracing argument in 1975 [26]: he wrote

$$v = G(u) = v_0 - \frac{2M\, B}{A} e^{-u/4M} \tag{3.23}$$

which is easily obtained from the above equation (again A and B are constants).

The explicit calculation performed here for the special case of a contracting ball allowed for reproducing the (more general) result obtained by Hawking when considering the behaviour of modes of a field in the vicinity of a black hole [25, 26]. In deriving Eq. (3.22), we have used the history of modes of the field passing through the gravitational field created by a collapsing ball of matter. Intuitively, we know that as the incoming waves propagate towards the shell of this ball, they will suffer a blueshift. Upon re-emerging from the ball and propagating out to J^+, they will be redshifted (this was anticipated by the earlier derivation of Sect. 2.1.2). Since we consider the case of a collapsing ball—that shrinks as the waves transit through it—the relative increase in the surface gravity experienced by the outgoing waves (with respect to that experienced by the incoming waves) will result in this redshift to be of exponentially larger amplitude than the blueshift.

From Eq. (3.22), we see that the incoming wave $f_\omega = e^{-i\omega v}$ is converted by the collapsing ball to the outgoing wave $F_\omega = e^{-i\omega - 4M \ln(\frac{v_0 - v}{2M\,B/A})}$. This factor $-4M \ln(\frac{v_0 - v}{2M\,B/A})$ represents the experience of an asymptotic observer at late time u: the outgoing null rays suffer an exponentially increasing redshift. Birrell and Davies pointed out that this redshift is the same as that of the surface luminosity of the collapsing ball (see [3, 27] and the derivation of 2.1.2.).

Note that the logarithmic dependence which governs the asymptotic form of the F modes on J^+ does not depend on the details of the metric *inside* of the ball. Indeed, it appeared in the last step of the matching sequence—using a flat spacetime metric for the interior of the ball was a mere mathematical trick that allowed for arriving at $g(v)$ ($G(u)$) without the complicated calculus presented in [3]. Hawking showed that the result at which we arrived here is more general than considerations of a ball or ball with a thin shell [25, 26].

3.1.2.2 Hawking Radiation

We will now build on the results of the above paragraph to show how black holes emit radiation. Essentially, the above considerations resulted in showing that a time-dependent background geometry would redshift waves propagating through a body collapsing to a black hole. In what follows, it will be shown that modes of a field that would be devoid of particles in a remote past would, after having propagated through the collapse, be sensed as populated by particles by an inertial detector in a remote future—this is Hawking radiation and results from the disruption of the modes of the quantum field as they propagate through the collapsing body. We will continue with the geometrical ray optics argument laid out previously, and follow the treatment of Hawking [25, 26], Parker [28, 29] and Davies as presented in Birrell and Davies' 1981 book [3]; again we shall digress slightly from their exact derivation but eventually arrive at the historical result. In doing so, we will introduce the fundamental tools and methods of quantum field theory in curved spacetime that will be used throughout this chapter, and to establish the novel results of this Thesis at a later stage.

We consider the massless scalar field ϕ, that obeys the wave equation

$$g^{\mu\nu}\Delta_\mu\Delta_\nu\phi = (-g)^{-1/2}\partial_\mu[(-g)^{1/2}g^{\mu\nu}\partial_\nu\phi] \tag{3.24}$$

in the Schwarzschild spacetime. Mode solutions of this equation are a complicated product of spherical harmonics and radial functions, but their detailed form is irrelevant to the present considerations—thus for the sake of simplicity we will resort to the f_ω and F_ω modes as defined in (3.11), as solutions to (3.24) in the *in* and *out* region respectively. We can do so if we remember that the dependence upon the angular coordinates must be the same for each term of all equations in this section. Following the general quantum theory of fields in curved spacetime [3] we decompose ϕ into a complete set of f_ω (positive frequency on $\mathcal{F}^-$) modes:

$$\phi = \int d\omega(a_\omega f_\omega + a_\omega^\dagger f_\omega^*). \tag{3.25}$$

The f_ω modes are normalized according to the condition

$$(f_{\omega_1}, f_{\omega_2}) = \delta_{\omega_1\omega_2}. \tag{3.26}$$

We assume that no scalar particles were present before the collapse began—the f_ω modes are in the quantum vacuum state

$$a_\omega|0\rangle = 0, \ \forall\,\omega. \tag{3.27}$$

We use the Heisenberg picture to study quantum states that span a Hilbert space—we will henceforth use the Fock representation as a basis in this space, thus identifying a_ω as the annihilation operator and $a_\omega^\dagger$ as the creation operator for quanta in the mode ω. See, for example, Chap. 2 in [3] for details of the quantization method on a curved background.

From Eq. (3.22), we can determine the form of the outgoing modes when traced back to $\mathcal{F}^-$

$$F_\omega \approx \begin{cases} e^{i\omega 4M\ln((v_0-v)/(2M\,B/A))}, & v < v_0 \\ 0, & v > v_0. \end{cases} \tag{3.28}$$

These outgoing modes are a complete orthonormal set of modes of the field ϕ, which may also be expanded in this set

$$\phi = \int d\omega(\bar{a}_\omega F_\omega + \bar{a}_\omega^\dagger F_\omega^*), \tag{3.29}$$

thus defining a new vacuum state $|\bar{0}\rangle$

$$\bar{a}_\omega|\bar{0}\rangle = 0, \ \forall\,\omega. \tag{3.30}$$

We expand the out modes in terms of the in modes

$$F_\omega = \int d\omega' (\alpha_{\omega\omega'} f_{\omega'} + \beta_{\omega\omega'} f_{\omega'}^*). \tag{3.31}$$

We can evaluate the $\alpha_{\omega\omega'}$, $\beta_{\omega\omega'}$ matrices by using Eqs. (3.31) and (3.26)

$$\alpha_{\omega\omega'} = (F_\omega, f_{\omega'}), \quad \beta_{\omega\omega'} = -(F_\omega, f_{\omega'}^*). \tag{3.32}$$

These matrices coefficients have the properties

$$\sum_k (\alpha_{\omega k} \alpha_{\omega' k}^* - \beta_{\omega k} \beta_{\omega' k}^*) = \delta_{\omega\omega'}, \tag{3.33}$$

$$\sum_k (\alpha_{\omega k} \beta_{\omega' k} - \beta_{\omega k} \alpha_{\omega' k}) = 0. \tag{3.34}$$

Note that one could also write the converse to (3.31), that is

$$f_\omega = \int d\omega (\alpha_{\omega\omega'}^* F_\omega + \beta_{\omega\omega'} F_\omega^*). \tag{3.35}$$

Equating the field expansions (3.25) and (3.29) and using (3.35), (3.31) together with the orthonormality of the modes (3.26) we can work out the relation between the annihilation operators attached to the incoming and outgoing modes—the Bogoljubov transformations [30]

$$a_{\omega'} = \sum_\omega (\alpha_{\omega\omega'} \bar{a}_\omega + \beta_{\omega\omega'}^* \bar{a}_\omega^\dagger), \tag{3.36}$$

and

$$\bar{a}_\omega = \sum_{\omega'} (\alpha_{\omega'\omega}^* a_{\omega'} - \beta_{\omega'\omega}^* a_{\omega'}^\dagger). \tag{3.37}$$

Glancing at (3.36), one remarks that the two Fock spaces based on the ingoing and outgoing modes are nontrivially different providing that $\beta_{\omega\omega'} \neq 0$. Thus the *in* vacuum state is not annihilated by the *out* annihilation operator (and vice versa). In fact the vacuum of the *out* modes contains a certain amount of particles in the *in* mode, as will be derived now.

We now rearrange the linear expansion (3.31) to identify the Bogoljubov coefficients, and insert the form of F on $\mathcal{F}^-$ (3.28),

$$\alpha_{\omega'\omega}^* = \frac{1}{2\pi} \sqrt{\frac{\omega'}{\omega}} \int_{-\infty}^{v_0} dv e^{i\omega' v} e^{i\omega 4M \ln((v_0 - v)/(2M \ B/A))}, \tag{3.38}$$

and

$$\beta^*_{\omega'\omega} = -\frac{1}{2\pi}\sqrt{\frac{\omega'}{\omega}}\int_{-\infty}^{v_0} dv e^{i\omega'v} e^{i\omega 4M \ln((v_0-v)/(2M\ B/A))}. \tag{3.39}$$

Both integrands are analytic everywhere except on the negative real axis, because of the branch cut of the logarithm function. Thus, and posing $v' = v_0 - v$,

$$\oint_C dv' e^{i\omega'v} e^{i\omega 4M \ln((v')/(2M\ B/A))} = 0 \tag{3.40}$$

around the closed contour C—which is a half-circle. Equivalently to Eqs. (3.38) and (3.39), we could have written in v':

$$\alpha^*_{\omega'\omega} = \frac{1}{2\pi}\sqrt{\frac{\omega'}{\omega}} e^{-i\omega v_0}\int_0^\infty dv' e^{-i\omega'v} e^{i\omega 4M \ln(v'/(2M\ B/A))}, \tag{3.41}$$

or

$$\beta_{\omega'\omega} = -\frac{1}{2\pi}\sqrt{\frac{\omega'}{\omega}} e^{-i\omega v_0}\int_0^\infty dv' e^{i\omega'v} e^{i\omega 4M \ln(v'/(2M\ B/A))}, \tag{3.42}$$

which are also analytic according to (3.40). Equating (3.38) and (3.41) and using a change of variables $v' \to -v'$ yields

$$\int_0^\infty dv' e^{i\omega'v} e^{i\omega 4M \ln(v'/(2M\ B/A))} = \int_0^\infty dv' e^{-i\omega'v} e^{i\omega 4M \ln(-v'/(2M\ B/A)-i\epsilon)}$$

$$= e^{4\pi M\omega}\int_0^\infty dv' e^{-i\omega'v} e^{i\omega 4M \ln(v'/(2M\ B/A))} \tag{3.43}$$

To arrive at this result, the relation $\ln(-v'/(2M\ B/A) - i\epsilon) = -i\pi + \ln(v'/(2M\ B/A))$ was used, with ϵ an infinitesimal variation in the phase introduced to clearly identify the relation between the norm Bogoljubov coefficients.[12] This relation is found by comparing (3.43) with Eqs. (3.41) and (3.42):

$$|\alpha_{\omega'\omega}| = e^{4\pi M\omega}|\beta_{\omega'\omega}|. \tag{3.44}$$

Earlier, the F modes were constructed as a set of positive frequency modes. But, as can be seen from Eq. (3.35) they are not a linear combination of the *in* f modes only: indeed, they are also expanded over some f^* modes. These have negative frequency with respect to the timelike Killing vector field in reference to which the F modes have positive frequency. Thus the set of *in* and *out* modes do not have a common vacuum state: some $\beta_{\omega'\omega}$ will be non zero and the F modes will contain a

---

[12]The calculation (3.38)–(3.43) was historically performed by means of Γ-functions [3, 26].

mixture of positive-(f) and negative-(f^*) frequency modes.[13]

We can find the mean number of particles created into mode ω by calculating the expectation value operator $N_\omega = a_\omega^\dagger a_\omega$ for the number of f-mode particles in the state $|\bar{0}\rangle$

$$\langle \bar{0}|N_\omega|\bar{0}\rangle = \sum_{\omega'} |\beta_{\omega'\omega}|^2 \neq 0. \tag{3.46}$$

That is, the vacuum of the $F_{\omega'}$ modes contains $\sum_{\omega'} |\beta_{\omega'\omega}|^2$ particles in the f_ω mode. In the present case this is

$$< N_\omega >= \frac{1}{e^{8\pi M\omega} - 1}, \tag{3.47}$$

a Planck spectrum with a temperature of

$$T_H = \frac{1}{8\pi M}, \tag{3.48}$$

the Hawking temperature of the black hole.

By Eq. (3.5), the temperature (3.48) is identical to (3.10) (with $\xi = 8\pi^2$), which we heuristically arrived at earlier—this demonstrates the thermodynamics basis of black holes. Black holes emit Hawking radiation, quanta spontaneously created from the vacuum that propagate away from it and can be observed at late times by an observer sitting away from the hole.

3.1.2.3   Origin of the Flux: Black Hole Evaporation

We have now resolved the paradox of black hole heat: quanta are emitted from the vacuum because of the disruption caused by the gravitational disturbance of imploding matter. This emission is thermal. Because it emits, the black hole can be in thermal equilibrium. Yet, this discovery raises a few questions: where are the particles emitted from, what is the source of this radiation, how does this process obey causality and conservation of energy, what would a freely falling observer see as they approach and cross the event horizon, what would an observer at late time see if looking at the black hole, how does Hawking radiation fit in the picture of black hole information, etc.

The purpose of this dissertation is not to present a Thesis that would have contributed to elucidate any of these intriguing concepts. Nonetheless, for the sake of completeness, this section will summarise some elements of answer to these many

---

[13]Particles will be present because $|\bar{0}\rangle$ will not be annihilated by a_ω:

$$a_\omega|\bar{0}\rangle = \sum_{\omega'} \beta^*_{\omega'\omega}|\bar{1}\rangle_{\omega'} \neq 0. \tag{3.45}$$

questions—the curious reader is advised to read the literature to go beyond the basic arguments that will be laid out below (see for example [3]).

First, let us recall the argument of Sect. 3.1.2: the wavelength of the quanta is, upon emission, comparable with the size of the hole. It is, therefore, impossible to localise the *origin* of this emission to within one wavelength: the particle concept is only useful near $\mathcal{F}^+$—in the vicinity of the hole, the radiation wavelength being comparable with the spacetime curvature, the concept of locally-defined particles is not valid. In his 1975 paper, Hawking called upon the concept of continuous spontaneous creation of virtual pairs of particle and antiparticle around the black hole to explain the origin of the radiation at late times [26]. In this picture, strong tidal forces in the vicinity of the hole could prevent re-annihilation of the pair that would be separated by a distance of the same magnitude as their wavelength of emission (the size of the hole). This would allow for one of the peers to escape to $\mathcal{F}^+$ and carry positive energy away from the hole, thus contributing to the Hawking flux (3.47), whilst the partner would enter the hole on a timelike path of negative energy relative to J^+. Alternatively, Hawking also suggested that the escaping quanta could have tunnelled through the event horizon out to $\mathcal{F}^+$ [31]. These two competing explanations are still being debated by the community—we shall not lay the arguments of each party down here, for they are irrelevant to this Thesis, though introducing the concept of partner particle will prove to be helpful when studying spontaneous emission from an optical event horizon.

One of the main arguments against the interpretations presented by Hawking following on his discovery of the radiation is the ill-defined nature of particles near the horizon. So, although the mechanism of radiation remains a mystery, one could still try and find out the source of its energy. It was proposed to do so by calculating the energy and flux density of this radiation at various positions around the hole—both locally defined quantities. The mathematical complexity of such an endeavour goes far beyond the scope of the present work; moreover, people have already dwelt upon it at length (see, for example [2–4, 32]). Typically, one would calculate the expectation value of the stress-energy-momentum in the state of the initial vacuum $|0\rangle$—which results in showing that the space curvature around a massive body induces a static vacuum stress. Unruh, Davies and Fulling were the first who envisaged this vacuum stress as a cloud of *negative energy* surrounding the body [33]. In the near-horizon region, between about $r = 3M$ and the horizon, the density of the cloud would be about the same (and proportional to $1/M^2$—it would have the energy of one photon of wavelength of the order of M emitted per time period of the order of M). The density of the cloud would then drastically decrease at larger rs. In studying the stress-energy-momentum tensor expectation value, they established that the source of the energy detected at infinity would come from the gravitational field itself: the negative energy cloud surrounding the hole would have a comparable energy magnitude to Hawking radiation and would continuously stream towards the central singularity, thus steadily reducing the mass-energy of the black hole. So the energy of the thermal Hawking flux does not come from inside the hole (nothing can cross the event horizon to outer space) but is provided for by the mass of the hole that depletes, because it receives negative mass-energy from the incoming vacuum stream. Unruh, Davies and Fulling showed in their paper that this description of the sourcing of Hawking radiation energy satisfies both causality and the conservation of energy [33].

Now when it comes to determining what an observer would actually measure, it is important to define precisely what question is being asked: the motion of the observer, as well as their localisation, with respect to the hole must be specified. Thinking along these lines, one would end up asking two questions: what can an observer who is freely falling on the hole detect? and what would an observer sitting away from the hole (Wheeler's bookkeeper) detect at late times? To answer these questions, we consider an observer equipped with some apparatus that is able to measure the total stress-tensor components. A detailed mathematical treatment of the present questions in a two-dimensional model of a black hole was first provided in [33] and then explicated further in [3].

A freely-falling observer, falling from a finite distance from the hole, would need a finite proper time to reach the event horizon. And yet, as measured from infinity (in u, v coordinates) the free-fall time is infinite (because of the effect of gravitational time dilation, see Sect. 2.1.2). Therefore, from a distance (say for Wheeler's book-keeper), the black hole will emit an infinite amount of radiation during the (infinite) time that the falling observer needs to reach the horizon. One would thus think that, to Wheeler's bookkeeper, the freely-falling observer should really encounter all the particles emitted by the black hole. As was demonstrated in Sect. 2.1.2, the free-falling observer will also appear redder and redder to Wheeler's bookkeeper as they approach the event horizon, until they seem to remain frozen there with infinitely long wavelength (and are thus actually invisible). Wheeler's bookkeeper cannot see the freely-falling observer reach the event horizon—and because the event horizon is only a global construct [31], it will not be experienced as a physical barrier by the freely-falling observer (they would not notice that they are reaching and crossing it). As they approach the horizon, the freely-falling observer will be surrounded by particles that are "shorter and shorter" (because their wavelength, size, is "inversely red-shifted" as the horizon is approached). Again, the ill-definiteness of the notion of particle prevents us from assessing what they will detect (how would the observer make sense of a counter click for a particle of significantly different wavelength than the apparatus dimensions?). Thus, no operational distinction is possible between the energy fluxes of Hawking radiation and that due to the sweeping of the observer's through the negative energy cloud: an observer who crosses the event horizon would measure a finite energy density (because the two divergences cancel out), in rather small amounts [2].

In contrast, as we stated above and as shown by Eq. (3.47), from afar, Wheeler's bookkeeper will detect a thermal flux coming from the hole (without being able to trace its exact origin back). In deriving Eq. (3.47), we didn't account for the effect of the gravitational potential on the flux at late times. As it turns out, there will be some backscattering of particle off the spacetime curvature surrounding the hole: only a fraction of the emitted flux will be able to reach out to the asymptotically flat regions of spacetime at $\mathcal{F}^+$. This is expressed by introducing a notion of probability for particles created in a mode F to escape to infinity: Γ the grey-body factor. Wheeler's bookkeeper will thus only detect a flux that is a filtered Planck spectrum (although this is thermal—see [3]): the further from the black hole the bookkeeper is, the lower the temperature, and the lower the frequency of the outpropagating modes the lower their temperature.

In arriving at Hawking's seminal result, we have used geometrical optics and assumed that late time, on $\mathcal{F}^+$, rays would have a finite frequency. These late time rays originate from the propagation of vacuum modes from $\mathcal{F}^-$ through the collapsing spacetime, that are scattered by the gravitational potential of the hole. Regardless of the exact event in spacetime at which Hawking radiation is emitted (i.e., at the horizon or in the vicinity of the hole), the rays get extremely frequency shifted, according to (2.23), as they propagate through the collapse and out to $\mathcal{F}^+$. In Sect. 2.1.2, we showed that radiation emitted *at* the event horizon would be infinitely redshifted as it propagates out to $\mathcal{F}^+$. Of course, the argument that allowed us to arrive at Hawking radiation features aspects of this catastrophic redshift: (3.43) really means that *out* modes (of retarded coordinate u) will acquire a phase of $e^{4\pi M\omega}$ (with M the mass-energy of the hole) with respect to the *in* modes (of retarded coordinate v). For *out* rays to have finite frequencies, *in* rays coming from $\mathcal{F}^-$ would have to have TransPlanckian frequencies—infinitely short wavelength. This is, of course, unphysical. This observation casts some shadow upon the validity of the derivation itself. To date, this is one of the main objections to the phenomenon of Hawking radiation, and neither the theories of General Relativity or Quantum Mechanics have provided a definite answer to what is infamously known as the *Transplanckian problem*. This hints at some Physics beyond our understanding. Fortunately, in analogue systems, this TransPlanckian Problem does not arise, thanks to dispersion—refer back to Sect. 2.1.3 for initial comments on this, and see the conclusion of the next section (Sect. 3.2.4.4) for further comments in the scheme of optical horizon.

Finally, let us examine the effect of Hawking radiation on the black hole itself: as particles are being radiated away, the hole will loose mass—thus increasing the flux and accelerating the mass-energy depletion further. Eventually, a black hole on which no positive-energy particles would fall would evaporate (and explode,[14] as hinted by the title of Hawking's foundational paper—"Black hole explosions?") [25]. The relationship between Hawking radiation and the information content of the hole, as well as the final fate of all the information of the hole, remain matters of passionate debate to date. The present Thesis will not attempt to contribute to these debates. Indeed, the very ability of analogue horizon systems to answer such questions has not been clearly established. We will instead focus on shedding light on the mechanism of Hawking radiation, the spontaneous emission of light from the vacuum.

---

[14]Black hole explosion refers to the fact that the emission rate goes as $1/M^2$ so that for small holes this becomes very large, and the lifetime (which goes as M^3) becomes very small.

3.1.3 Conclusion and Discussion

In this part of the dissertation, we have used the tools developed in the early 1970s to investigate the then paradoxical black hole heat. We have established that a stationary black hole would disturb modes of a massless scalar field in such a way that, when propagating from remote past infinity to remote future infinity through the gravitational collapse, they would be extensively redshifted. Furthermore, we have shown how this disruption, caused by the gravitational disturbance of imploding matter, would result in field quanta to be emitted in a thermal flux propagating from the hole out to infinity. This, we found, was due to the relationship between incoming and outgoing modes in our field theory: because they do not span the same vector space (essentially the gravitational disturbance can be seen as an impedance mismatch between the *in* and *out* regions) positive and negative frequency *in* modes mix in forming positive frequency *out* modes. We then digressed from our mathematical path to dwell upon considerations that are still under discussion regarding this thermal flux, Hawking radiation. In particular, we introduced the concept of pairs of particles and of negative energy falling on the hole to explain what the source of the Hawking radiation energy flux is.

In what follows, we will see how such ideas can be ported to the experimental scheme of fibre analogue event horizon (as proposed by the St Andrews collaboration in 2008 [34]). We will use a model for light and matter interaction in a dispersive medium to establish a wave equation analogous to the Eddington-Finkelstein metric by following the method laid out by Unruh (see Ref. [35] and Sect. 2.1.3 for details of the method). This will reveal how light scatters from negative to positive frequency modes (and vice-versa) at the horizon—which is the essence of the Hawking emission mechanism—and leads us to the experimental idea that would allow for proving the reality of the Hawking emission mechanism.

3.2 Quantum Field Theory in a Condensed Matter System

3.2.1 Quantum Vacuum Emission from a Refractive Index Front

3.2.1.1 Rational for a Theory

In the preceding section, we have used Hawking's semi-classical theory of fields in curved spacetime background: we have studied the fate of modes of a quantised fields as they propagate in a classical and evolving medium. We thus derived his 1974 result [25] that black holes emit a thermal flux, Hawking radiation (HR). Glancing back at the final result of Sect. 3.1.2.2, that HR is characterized by a Planck spectrum

(3.48), and expressing it in standard units, we can understand why it has never been detected:

$$T_{Hmax} = \frac{\hbar c^3}{8\pi G M k_B} \approx \frac{1.227 \times 10^{23} kg}{M} K = 6.169 \times 10^{-8} \frac{M_S}{M} K. \qquad (3.49)$$

If we insert the mass of the lightest black hole possible [9], $M_{BH} \approx 1.39 M_S$, in (3.49) we obtain $T_{Hmax} = 85.75$ nK. This is 8 orders of magnitudes below the colour temperature of decoupled photons that form the Cosmic Microwave Background (presently of about 2.7260 K [36]). In other words, HR is hidden from us by the universe's own glow. It is a euphemism to state that Hawking radiation is *hard* to see.

Fortunately, as we saw in Chap. 2, Unruh showed [35] that, in total analogy with their astrophysical counterparts, dumb holes should emit a thermal flux. Beyond the realisation that the kinematics were analogous, Unruh's crucial insight was that once the analogy has been drawn, it is possible to repeat Hawking's 1974 semi-classical argument, only replacing light with perturbations in the field under study (*eg* the acoustic field in [35]) to arrive at the conclusion that analogue horizons emit quanta from the vacuum. He predicted that quantum hydrodynamical fluctuations in a moving fluid would convert into pairs of phonons at the sonic horizon—thus reviving the hopes to at least shed light on the Hawking emission mechanism. Note that in this Thesis, we are agnostic about the identity of quantum vacuum emission from analogues, i.e., we do not claim or disclaim that it is HR.

Following on the 2008 seminal experimental demonstration of the realisation of an analogue horizon in optical fibres, in which the authors predicted that the moving horizons would spontaneously radiate a thermal flux of a 1000 K [34], a handful of groups have assembled optical analogue experiments: Faccio in Heriot-Watt in the UK and previously at Insubria, Como, Italy [37], Leonhardt at the Weizmenn Institute in Israel [38], Genty and Murdoch in Tempere, Finland, and in Aukland, New Zealand [39], but none of them has managed to observe the spontaneous emission of light from the vacuum. This is partly due to the lack of detailed analytical predictions of the wavelength and intensity of the radiation for an actual experiment (although a wealth of numerical studies has been carried out by Faccio's group, see for example [40–42], and others [39, 43–45]). Moreover, the role and influence of dispersion in the details of the mechanism of spontaneous emission remains a topic of active study. For example, in recent works, other authors [46–48] have calculated the Hawking temperature T_H from the surface gravity $\frac{\partial_\zeta g_{00}}{g_{01}}|_{horizon}$ at the analogue horizon for various Refractive Index Front (RIF) profiles (smooth and abrupt variations in the refractive index) in dispersionless media. In addition, Unruh and collaborators have discussed the rise of a grey body factor under the breakdown of conformal invariance in a similar analogue toy model as Eq. (2.79) in [49].

Of course, the geometry of the RIF is an important factor in the ability to study the characteristics of the spontaneous emission: for example, in the case of a pulse in an optical fibre, the length of the pulse front has to be comparable to the wavelength of radiation [34]. Furthermore, only smooth variations in the refractive index can be studied if one wishes to address concepts such as the relation between the temperature of spontaneous emission and the surface gravity (a step function in the refractive index formally corresponds to infinite surface gravity[15]). And yet, understanding the critical conditions needed to observe the spontaneous emission of photons at an optical horizon can be greatly helped by analytically studying a step-like RIF geometry[16] [47, 48, 50–52].

In particular, Finazzi and Carusotto developed a fully quantised analytic 1 + 1D model based on a sharp step behaviour of the dielectric properties of a nonlinear medium in [47]. At this moving boundary (RIF) between two multibranch dispersive media, certain modes may experience either analogue black- or white-hole or horizonless configurations, leading to mode mixing and spontaneous emission of radiation. In all configurations, the mismatch in the medium properties on either side of the boundary leads to the mixing of modes of opposite norm and thus to spontaneous emission of radiation. They performed numerical evaluations (based on the material properties of fused silica) of the pair-production processes involved and discovered that emission is dominant over optical frequencies. In their studies [47, 51], they focused only on emission spectra in positive-norm optical modes of light. However, emission from the vacuum always comes in a pair of positive-and negative-norm modes. As is exemplified by the existence of a (negative energy- or frequency) partner particle to HR, the negative norm modes of the theory play the role of the partner mode in the Hawking emission mechanism (at the output, one obtains a two-modes squeezed state). This is relevant, in particular, because these negative norm modes emit at different laboratory frequencies than their positive-norm parter modes. Besides, in order to maintain these different configurations, Finazzi and Carusotto finely adapted the velocity of the RIF when changing the nonlinearity. However, the nonlinearity in the experiment typically changes independent of the RIF velocity, leading to a spectral structure strongly dependent on the nonlinearity strength, as well as to a scaling of the signal with nonlinearity.

3.2.1.2 Outline of the Theory

The results that will now be presented—that form the theoretical component of this Thesis—were published in the summer of 2015 in *Physical Review A*, see [53]. We use the model [47] to reveal the above-mentioned properties of quantum vacuum emission by following the steps outlined here:

- we first expand the analytical model to consider emission from all modes of all norms at any frequency and change of refractive index;

---

[15]Although a step-like profile models an infinite slope at the horizon, which would correspond to an infinite surface gravity and temperature, the calculations show a totally different result. As we will see, the spectral densities we calculate are finite. I think this is because, ultimately, the amplitude of waves is limited by dispersion.

[16]In the experiment only smooth profiles can be realised. Calculations with an infinitely steep profile only have a suggestive role in understanding the experiment.

- hence we obtain emission for different refractive index changes in the frame co-moving with the RIF without tuning the RIF velocity;
- we convert the spectra to the laboratory frame, including all mode contributions, inclusive of the important negative-norm ones;
- finally, we find the scaling law for the total photon flux associated with black-hole emission with increasing nonlinearity—RIF height.

These will all be essential in identifying emission in future optical event horizon experiments. We will also show how these results yielded the intuition behind the experiment that we performed in 2016–2017 and that will be presented in the final chapter of this dissertation.

We begin with an introduction of the theoretical model of the scattering of vacuum modes at the horizon. We detail how the interaction of light and matter in a uniform dispersive medium is modelled, and identify the eigenmodes and study their properties. We then extend this model to consider an inhomogeneous medium composed of two distinct homogeneous regions (of different optical properties) separated by a moving RIF. We proceed to constructing eigenmodes of this nonuniform medium and to describing the scattering of these eigenmodes, that is the mode conversion process at the RIF, by the Scattering Matrix formalism. Finally, we quantize the field modes and calculate the photon flux density in the moving and laboratory frames. Next, in Chap. 4, we consider light-matter interactions in bulk silica—and compute the spectra of emission in both frames. These spectra allow us to identify in detail the contributions of the various modes to the emission, and the role of analogue event horizons. We also integrate the spectra to evaluate the total emission and its dependence on the refractive-index height.

3.2.2   Light-Matter Interactions in a Dispersive Medium

3.2.2.1   Lagrangian Electrodynamics

In this subsection, we lay out the field theory model that will later support the theoretical framework of scattering at a Refractive Index Front (RIF). We begin with considerations drawn from Electrodynamics, that is the classical description of the dynamics of the total system (in the present case, the electromagnetic field in a non-relativistic medium). This will later enable us to describe the interaction processes between radiation and matter: the scattering of field modes and emission of photons.

In order to describe the interactions of light with a homogeneous and transparent dielectric medium, we employ a microscopic model inspired by the Hopfield model of Condensend Matter Theory [54], as was first suggested by Schützhold and collaborators in [55]. We restrict ourselves to a one-dimensional geometry and scalar electromagnetic fields and operate at frequencies sufficiently far from the medium resonances to neglect absorption. Matter, in the model, consists of polarisable molecules, harmonic oscillators of eigenfrequency (resonant frequency) $\Omega_i = \frac{2\pi c}{\lambda_i}$ and elastic constant κ_i^{-1}. In the medium, there is one such harmonic oscillator at each point in space, but since the coupled electromagnetic field has a large wavelength compared to the molecular scale of the dielectric, we can consider the dielectric in the continuum limit[17] and describe the electric dipole displacement by the massive scalar field P_i. The electromagnetic field (a massless scalar field) in the medium is described by the vector potential $A(x, t)$ via $\boldsymbol{E} = -\partial_t \boldsymbol{A}$ in temporal gauge.[18] In order to reproduce the refractive index of most materials, we shall henceforth consider a medium featuring three resonances. In the rest frame of the medium—the laboratory frame—the interaction of the electromagnetic field with the three polarization fields of the medium is described by the Lagrangian density [47, 54, 56]

$$\mathcal{L}_{LF} = \frac{(\partial_t A)^2}{8\pi c^2} - \frac{(\partial_x A)^2}{8\pi} + \sum_{i=1}^{3} \left(\frac{(\partial_t P_i)^2 \lambda_i^2}{2\kappa_i (2\pi c)^2} - \frac{P_i^2}{2\kappa_i} + \frac{A}{c} \partial_t P_i \right) \qquad (3.50)$$

where the inertia of the harmonic oscillators P_i when subjected to an external drive is $\frac{\lambda_i^2}{\kappa_i (2\pi c)^2}$. The term linear in A in Eq. (3.50) describes the coupling between the fields. The Lagrangian density accounts for the free space and medium contributions to the field through the first two terms and the sum, respectively. Dispersion enters as a time dependence of the addends of the summation.

From the strong coupling of light with the polar excitations of the medium result polaritons—hybrid light and matter quasiparticles [57]. As is illustrated on Fig. 3.2, the coupling of the photon with the excitons leads to an energy anticrossing of the bare oscillators, thus giving rise to new normal modes of the system known as polariton branches. The energy shift depends on the overlaps of the electromagnetic field and polarisation fields; it is proportional to the coupling constant κ_i^{-1}. In the case of a medium with three resonances, the dispersion relation resulting from the anticrossing features four branches arranged around 3 poles: the "top" branch exhibits gradient larger than the speed of light in the medium whilst the lower (energy) branches, labelled as "upper", "middle", and "lower", are characterized by a non-parabolic energy-momentum dispersion. Hereafter, we will only study frequencies over which the effects attached to the "top" branch can be neglected. Then, the non-parabolic behaviour of the remaining 3 dispersion branches leads to the *effective-mass* approximation [58] according to which polaritons have an effective mass and inherit from excitons the capacity to interact with each other.[19] Hence, Eq. (3.50) describes a massive scalar field whose modes of oscillation can couple to each other—in what follows we will study such coupling when those modes scatter at a boundary between two regions of different refractive index.

---

[17]The model does not account for the dispersion changes due to the finiteness of the intersites distance.

[18]Note that in this section, the partial derivative with respect to a variable is denoted by $\partial_t \equiv \frac{\partial}{\partial t}$. We do not use the relativistic-covariant formulation.

[19]Note that the lowest branch is approximately a massless polariton: it can be fitted with a dispersion relation of the form $|\omega| = c|k|$ for low wavenumbers (close to $k = 0$).

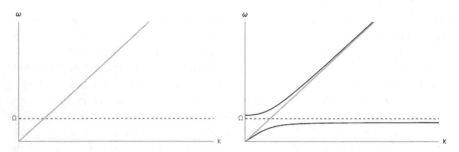

Fig. 3.2 Light-matter interaction in a dispersive medium. A photon, represented by the green contour line (of gradient the speed of light c) interacts with the exciton of a medium (represented by the dotted red line, set at the resonant frequency of the medium Ω). This interaction results in the apparition of two distinct polaritons. These quasi particles obey the dispersion relation. The interaction of one electromagnetic field with one polarisation field yields a two-branches dispersion relation. The shape of the branches, and distance between them at the closest point (anticrossing), depend on the elastic constant κ^{-1} of the exciton of the medium via the inertia $(\kappa\Omega^2)^{-1}$. The lower polariton branch asymptotically tends to Ω, whilst the upper polariton branch asymptotically tends to c

3.2.2.2 Action

The step in refractive index (RIF) is propagating in the positive x direction at speed u. It is convenient to express the Lagrangian density (3.50) in a frame co-moving with the RIF by applying a Lorentz boost

$$\begin{pmatrix} x \\ t \end{pmatrix} = \gamma \begin{pmatrix} 1 & u \\ \frac{u}{c^2} & 1 \end{pmatrix} \begin{pmatrix} \zeta \\ \tau \end{pmatrix}. \tag{3.51}$$

In this moving frame, the system is stationary—the medium properties are independent of time. The Lagrangian density for light in a homogeneous medium there reads

$$\mathcal{L}_{MF} = \frac{(\partial_\tau A)^2}{8\pi c^2} - \frac{(\partial_\zeta A)^2}{8\pi} +$$
$$\sum_{i=1}^{3} \left(\frac{\gamma^2 (\partial_\tau P_i - u\partial_\zeta P_i)^2 \lambda_i^2}{2\kappa_i (2\pi c)^2} - \frac{P_i^2}{2\kappa_i} + \frac{A\gamma}{c}(\partial_\tau P_i - u\partial_\zeta P_i) \right). \tag{3.52}$$

See (2.81) for the transformation of the differentials. By the principle of least action [59], we obtain the Hamiltonian density by varying the Lagrangian density (3.52) with respect to the canonical momentum densities of light

$$\Pi_A = \frac{\partial \mathcal{L}_{MF}}{\partial (\partial_\tau A)} = \frac{\partial_\tau A}{4\pi c^2} \tag{3.53}$$

and polarisation fields

$$\Pi_{P_i} = \frac{\partial \mathcal{L}_{MF}}{\partial(\partial_\tau P_i)} = \frac{\gamma^2 \lambda_i^2 (\partial_\tau P_i - u \partial_\zeta P_i)}{\kappa_i (2\pi c)^2} + \frac{A\gamma}{c}. \tag{3.54}$$

Thus

$$\mathcal{H}_{MF} = \frac{1}{2} \left(\partial_\tau A \Pi_A + \Pi_A \partial_\tau A + \sum_{i=1}^{3} \left(\partial_\tau P_i \Pi_{P_i} + \Pi_{P_i} \partial_\tau P_i \right) \right) - \mathcal{L}_{MF}. \tag{3.55}$$

From the Hamiltonian density follow the Hamilton equations, the equations of motion for the fields [59]:

$$\begin{cases} \partial_\tau \psi_j = \frac{\partial \mathcal{H}}{\partial \Pi_j} \\ \partial_\tau \Pi_j = -\frac{\partial \mathcal{H}}{\partial \psi_j} + \sum_i \partial_i \frac{\partial \mathcal{H}}{\partial(\partial_i \psi_j)} \end{cases} \tag{3.56}$$

where ψ_j and Π_j are any of the field and conjugate momenta A, P_i, Π_A or Π_{P_i}, respectively. We complexify the massive field obtained from the action of (3.52) by demanding plane wave solutions of the form

$$V = V_{\omega'} e^{ik'\zeta - i\omega'\tau}, \tag{3.57}$$

where V is the eight-dimensional field operator $V = (A\ P_1\ P_2\ P_3\ \Pi_A\ \Pi_{P_1}\ \Pi_{P_2}\ \Pi_{P_3})^T$, to the dynamical equations (3.56). In Fourier space, $\partial_\tau = -i\omega'$ and $\partial_\zeta = ik'$, and (3.56) reads

$$\begin{cases} -i\omega' A = 4\pi c^2 \Pi_A, \\ -i\omega' P_i = \frac{\kappa_i (2\pi c)^2}{\gamma^2 \lambda_i^2} \left(\Pi_{P_i} - \frac{A\gamma}{c} \right) + u i k' P_i, \\ -i\omega' \Pi_A = -\frac{k'^2 A}{4\pi} + \sum_{i=1}^{3} \left(\frac{\kappa_i (2\pi c)^2}{\gamma^2 \lambda_i^2} \left(\Pi_{P_i} - \frac{A\gamma}{c} \right) \right), \\ -i\omega' \Pi_{P_i} = -\frac{P_i}{\kappa_i} + u i k' \Pi_{P_i}. \end{cases} \tag{3.58}$$

Eliminating the fields in (3.58), simple algebra then leads to the generic Sellmeier dispersion relation of bulk transparent dielectrics:

$$c^2 k^2 = \omega^2 \left(1 + \sum_{i=1}^{3} \frac{4\pi \kappa_i}{1 - \frac{\omega^2 \lambda_i^2}{(2\pi c)^2}} \right) \tag{3.59}$$

where the Lorentz transformations from the laboratory frame to the moving frame was used to identify $\omega' = \gamma(\omega - uk)$ and $k' = \gamma(k - \frac{u}{c^2}\omega)$. This dispersion relation[20] is

---

[20]Note that (2.78) is an approximate version of this dispersion relation where we have assumed that $\omega < |\Omega|$ for a medium with only one resonant frequency.

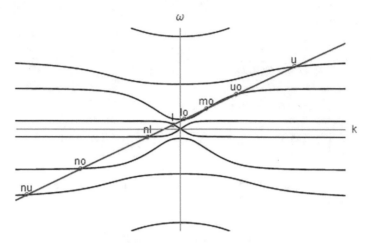

Fig. 3.3 Sellmeier dispersion relation, Eq. (3.59), with three resonances in the laboratory frame. There are eight branches (black lines). A contour of ω' is shown in blue. Their intersection points indicate the modes of propagation in the medium (red circles)

plotted in Fig. 3.3: there are eight branches, four with positive laboratory frequency, and their four negative laboratory frequency counterparts, symmetric about the k axis.

3.2.2.3 Noether's Theorem and Norm by the Scalar Product

By construction, the complexified Lagrangian

$$
\mathcal{L}_{MF}^{\text{complex}} = \frac{1}{2} \left(\frac{\partial_\tau A^* \partial_\tau A}{4\pi c^2} - \frac{\partial_\zeta A^* \partial_\zeta A}{4\pi} \right) +
$$

$$
\frac{1}{2} \sum_{i=1}^{3} \left(\gamma^2 \frac{\lambda_i^2}{\kappa_i (2\pi c)^2} (\partial_\tau P_i^* - u\partial_\zeta P_i^*)(\partial_\tau P_i - u\partial_\zeta P_i) - \frac{|P_i|^2}{\kappa_i} \right. \tag{3.60}
$$

$$
\left. + \gamma \frac{A}{c} (\partial_\tau P_i^* - u\partial_\zeta P_i^*) + \gamma \frac{A^*}{c} (\partial_\tau P_i - u\partial_\zeta P_i) \right)
$$

is invariant under any transformation of the global phase of the dynamic fields ($A \rightarrow e^{i\phi} A$ and $P_i \rightarrow e^{i\phi} P_i$, likewise for the complex conjugate fields) [48, 49]. From the Lagrangian (3.60), one can calculate the charge density (the net charge per unit volume) [59]

$$
\rho = i(\Pi_A^* A + \sum_{i=1}^{3} \Pi_{P_i}^* P_i - \sum_{i=1}^{3} P_i^* \Pi_{P_i} - A^* \Pi_A) \tag{3.61}
$$

as well as the current density, the rate of change of charge over time per unit length dl,

$$j = \int_V dl \frac{\partial \rho}{\partial \tau} = -iu(\Pi_A^* A + \sum_{i=1}^{3} \Pi_{P_i}^* P_i - \sum_{i=1}^{3} P_i^* \Pi_{P_i} - A^* \Pi_A). \qquad (3.62)$$

According to Noether's theorem [60], the continuous symmetry of $\mathcal{L}_{MF}^{complex}$ (3.60) implies a conserved current $\partial_\tau \rho + \partial_\zeta j = 0$—this is the *continuity equation*. In the moving frame, being the system stationary, $\partial_\tau \rho = 0$. Thus the continuity equation for Noether's current simplifies to $\partial_\zeta j = 0$: the current density is a space-time-independent quantity for the plane wave modes (3.57). Integrating Noether's charge density (3.61) results in nothing else than calculating the product of the field operator V with itself, $\rho = iV^\dagger \eta V$:

$$\int_{-\infty}^{\infty} \rho d\zeta = \langle V, V \rangle. \qquad (3.63)$$

η is the symplectic (or selection) matrix—$\eta = \begin{pmatrix} 0 & I_4 \\ -I_4 & 0 \end{pmatrix}$, with I_4 the 4×4 identity matrix. Equation (3.63) is the norm of an eigenmode[21]—an orthonormal plane-wave solution—of the system by the conserved scalar product

$$\langle V_1, V_2 \rangle = \frac{i}{\hbar} \int d\zeta V_1^\dagger(\zeta, \tau) \eta V_2(\zeta, \tau)$$
$$= \frac{i}{\hbar} \int d\zeta \left(A_1^* \Pi_{A_2^*} - \Pi_{A_1} A_2 + \sum_{i=1}^{3} \left(P_{i,1}^* \Pi_{P_{i,2}^*} - \Pi_{P_{i,1}} P_{i,2} \right) \right) \qquad (3.64)$$

defined on the set of our Hamilton equations generalised to complex values (3.58). The $\hbar^{-1}$ prefactor was inserted for normalisation purposes.

As a result of our application of Noether's theorem—the space-time independence of the charge density—, this Klein-Gordon [3, 59] product (3.64) is a conserved quantity in τ, and therefore the norm of the state is conserved. The former can be formally proven by the following algebraic calculation [61]:

---

[21]Note that, by replacing the conjugate momenta of the electromagnetic and polarisation fields by their expression in terms of derivatives of the fields (Eqs. 3.53 and 3.54), one obtains the usual form of the pseudo norm—as in Eq. (1.12), with ϕ a field. Because of dispersion, this expression would of course be slightly more complicated, although as readily computable.

$$\partial_\tau \langle V_1, V_2 \rangle = \frac{i}{\hbar} \int d\zeta \partial_\tau \left(A_1^* \Pi_{A_2} - \Pi_{A_1}^* A_2 + \sum_{i=1}^{3} \left(P_{i,1}^* \Pi_{P_{i,2}} - \Pi_{P_{i,1}}^* P_{i,2} \right) \right)$$

$$= \frac{i}{\hbar} \int d\zeta \left(\partial_\tau A_1^* \Pi_{A_2} - \Pi_{A_1}^* \partial_\tau A_2 + A_1^* \partial_\tau \Pi_{A_2} - \partial_\tau \Pi_{A_1}^* A_2 \right.$$

$$\left. + \sum_{i=1}^{3} \left(\partial_\tau P_{i,1}^* \Pi_{P_{i,2}} - \Pi_{P_{i,1}}^* \partial_\tau P_{i,2} + P_{i,1}^* \partial_\tau \Pi_{P_{i,2}} - \partial_\tau \Pi_{P_{i,1}}^* P_{i,2} \right) \right)$$

$$(3.65)$$

The computation of the first terms yields

$$\text{1st term} = \frac{i}{\hbar} \int d\zeta \left(4\pi c^2 \Pi_{A_1}^* \Pi_{A_2} + \partial_\zeta A_1^* \Pi_{A_2} - \Pi_{A_1}^* 4\pi c^2 \Pi_{A_2} - \Pi_{A_1}^* \partial_\zeta A_2 \right.$$

$$+ A_1^* \frac{\partial_\zeta^2 A}{4\pi} + A_1^* \partial_\zeta \Pi_{A_2} + A_1^* \sum_{i=1}^{3} \frac{\kappa_i \Omega_i^2}{c} \left(\Pi_{P_{i,2}} - \frac{A_2}{c} \right)$$

$$\left. - A_2 \frac{\partial_\zeta^2 A_1^*}{4\pi} - A_2 \partial_\zeta \Pi_{A_1}^* - A_2 \sum_{i=1}^{3} \frac{\kappa_i \Omega_i^2}{c} \left(\Pi_{P_{i,1}}^* - \frac{A_1^*}{c} \right) \right)$$

$$= \frac{i}{\hbar} \int d\zeta \sum_{i=1}^{3} \frac{\kappa_i \Omega_i^2}{c} \left(A_1^* \Pi_{P_{i,2}} - \Pi_{P_{i,1}}^* A_2 \right),$$

$$(3.66)$$

and that of the addends of the summation

$$\text{2nd term} = \frac{i}{\hbar} \int d\zeta \sum_{i=1}^{3} \left(\kappa_i \Omega_i^2 \left(\Pi_{P_{i,1}}^* - \frac{A_1^*}{c} \right) \Pi_{P_{i,2}} + \partial_\zeta P_{i,1}^* \Pi_{P_{i,2}} - \Pi_{P_{i,1}}^* \partial_\zeta P_{i,2} \right.$$

$$\left. - \Pi_{P_{i,1}}^* \kappa_i \Omega_i^2 \left(\Pi_{P_{i,2}} - \frac{A_2}{c} \right) + P_{i,1}^* \frac{-P_{i,2}}{\kappa_i} - \frac{-P_{i,1}^*}{\kappa_i} P_{i,2} - \partial_\zeta \Pi_{P_{i,1}}^* P_{i,2} \right)$$

$$= \frac{i}{\hbar} \int d\zeta \sum_{i=1}^{3} \frac{\kappa_i \Omega_i^2}{c} \left(-A_1^* \Pi_{P_{i,2}} + \Pi_{P_{i,1}}^* A_2 \right).$$

$$(3.67)$$

Clearly, $\partial_\tau \langle V_1, V_2 \rangle = 0$.[22]

Note that the scalar product (3.64) is not positive definite, and thus the norm of all mode solutions—which is inherited from this scalar product—is not necessarily positive. In fact, modes that have a negative frequency in the laboratory frame have a negative norm, whilst modes that have a positive frequency in the laboratory frame

---

[22] An alternative proof follows from the observation that, given $\partial_\tau \rho = 0$ and $\langle V_1, V_2 \rangle = \alpha \langle V_1, V_1 \rangle + \sum_{i=1}^{3} \bar{\alpha}_i \langle V_i, V_i^\dagger \rangle$, being the second term of the latter equation zero, the assessment of time conservation consists in calculating $\partial_\tau \int \alpha \langle V_1, V_1 \rangle d\zeta + \partial_\tau \langle V_1, V_1 \rangle = \int \partial_\tau \alpha \langle V_1, V_1 \rangle d\zeta$. $\partial_\tau \alpha \langle V_1, V_1 \rangle = 0$, and thus $\partial_\tau \alpha \langle V_1, V_2 \rangle = 0$.

have a positive norm: Finazzi and Carusotto found that the sign of $\langle V, V \rangle$ depends upon that of

$$\frac{i}{\hbar} V_{\omega'}^{\alpha\dagger} \eta V_{\omega'}^{\alpha} = \gamma c^2 \frac{|C_{\omega'}|}{2\pi} \frac{k}{v_g} \left(1 - \frac{u v_g}{c^2}\right) \tag{3.68}$$

where $C_{\omega'}$ is a moving frame frequency dependent normalisation factor for the mode solutions [47]. Since $u < c$, the term in brackets in (3.68) is always positive, thus the sign of the scalar product of V with itself depends upon the ratio $\frac{k}{v_g}$. It is easy to find an expression for v_g, the group velocity of a mode solution in the laboratory frame from the dispersion relation (3.59)—by definition, $v_g = \frac{d\omega}{dk}$, and one calculates

$$\frac{dk}{d\omega} = \frac{\omega}{k} \frac{dk^2}{d\omega^2} = \frac{\omega}{c^2 k} \left(1 + \sum_{i=1}^{3} \frac{4\pi \kappa_i}{(1 - \omega^2/\Omega_i^2)^2}\right). \tag{3.69}$$

wherefrom, glancing back at (3.68), it is obvious that $\mathrm{sign}\|V\| \propto \mathrm{sign}(\omega)$ [47–49, 53, 61].

For simplicity, we shall henceforth refer to the *pseudo*-norm (3.63) as the norm of the field solution. Taking the complex conjugate of Eq. (3.63), $\langle V^*, V^* \rangle$, yields a result of opposite sign to the inner product of V with itself. Thus modes belonging to the upper (lower) half plane of the dispersion relation in energy momentum space Fig. 3.3 have positive (negative) norm. In the comoving frame, positive-frequency waves with negative norm appear. Such negative norm modes were recently observed in water wave experiments [62, 63] and in optics [64–66]. As we will see in a later section, positive moving-frame-frequency negative-norm modes are associated with spontaneous emission from the quantum vacuum. Due to the conservation of norm, the generation of negative-norm waves signifies a simultaneous increase in positive-norm waves, the generation of correlated waves.

3.2.3   Mode Configurations at a Refractive-Index Front

In the previous section we presented a canonical model aiming at describing the phenomenology of light and matter interaction in a dielectric medium. We found that mode solutions of the complexified fields equations of motion in a homogeneous medium could have positive or negative norm as a function of their frequency in the laboratory frame (the rest frame of the medium). We will now push our classical study of the electrodynamics of the system further to describe a non-uniform medium.

3.2.3.1   Phenomenology of the Refractive Index

In this Thesis, we consider the simple geometry of a RIF as shown in Fig. 3.4 in the comoving frame. The medium is composed of two homogeneous regions, separated by the RIF at $\zeta = 0$, creating a step in the refractive index. The boundary at $\zeta = 0$ constitutes an infinitely steep RIF which propagates in a steady and rigid way in the positive ζ direction. Phenomenologically, the refractive index of a homogeneous region is described by the dispersion relation (3.59), with dispersion parameters $\kappa_{i,R}$ ($\kappa_{i,L}$) and $\lambda_{i,R}$ ($\lambda_{i,L}$) in the right (left) region. The change in refractive index between the left and right regions is modelled by the step height δn, defined by

$$n(\zeta) = n_L \theta(-\zeta) + n_R \theta(\zeta) = n_R + \delta n \theta(-\zeta) \tag{3.70}$$

$\theta(\zeta)$ is the Heaviside step function; and illustrated in Fig. 3.4. In an extension of the oscillator model by Drude and Lorentz, the index change is described by the scaled Sellmeier coefficients

$$\begin{aligned} \kappa_{iL} &= \sigma \kappa_{iR} \\ \lambda_{iL}^2 &= \sigma \lambda_{iR}^2, \end{aligned} \tag{3.71}$$

where, for small index changes, it follows from (3.59) that

$$\sigma \approx 1 + \frac{2n_R \delta n}{n_R^2 - 1}, \tag{3.72}$$

with n_R is the refractive index on the right side [47, 53].

Note that the present microscopic model of the dielectric [54]—a set of harmonic oscillators (whose properties are described by the κ_i and λ_i position-dependent constants)—cannot possibly account for the reality of the medium. Instead, it is a simple, phenomenological means to obtain the dielectric constant of the medium [67]. Indeed, a linear dielectric constant results in reality from a nontrivial collection of quantum processes. These would be further complicated when considering a nonlinear dielectric. Therefore, the modulation (3.72) of both constants of the oscillators by (3.71) is merely a proposal to describe the change of the dielectric constant within a self-consistent theory. To me, this means that the details of the change in the dielectric constant can equally be accounted for by a modulation of both or either of κ_i and λ_i. For the sake of the present work, I have decided to change both (by (3.71))—others have proceeded likewise (see for example [47]) or otherwise (see for example [48, 49]).[23] A full review of the various approaches, as well as a thorough verification of the independence of the change in the dielectric constant on the details of the model, would be important and shall be the subject of future work (see Appendix B for additional details).

3.2.3.2 Modes in an Inhomogeneous Medium

We saw earlier (see 3.2.2.3) that, as a consequence of the continuous symmetry of the complexified Lagrangian (3.60), Noether's theorem yielded a space-time inde-

---

[23]Remark that the change in the refractive index described by (3.71) is frequency-dependent.

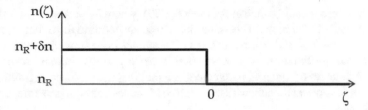

Fig. 3.4 Sketch of the RIF in the moving frame: there are two homogeneous regions of uniform refractive index on the left and right of a dielectric boundary of height δn

pendent current density for plane wave modes (3.57) of the field in a homogeneous region. This implies that energy is conserved in the moving frame, that is the comoving frame frequency ω' is a conserved quantity—this translates to the condition

$$\gamma(\omega - uk) = \text{const} \tag{3.73}$$

in the laboratory frame (see Refs. [49] or [44, 48] for other ways to arrive at this condition in different setups). This condition is a straight contour-line of slope u and ω-intercept $\omega = \omega'$ in Fig. 3.3. In terms of polariton physics, this means that we can identify the modes of propagation of the massive field subject to the dispersion relation in both regions for a given ω'. Thus solutions of fixed ω' are found at the intersection points between a line of constant ω' with the various polariton branches in the dispersion diagram (red circles in Fig. 3.3).

Combining Eqs. (3.73) and (3.59), the dispersion relation in the laboratory frame with the conservation of energy in the moving frame, yields the condition that mode-solutions to the equation of motion have to obey. The dispersion relation (3.59) is an eighth order polynomial, thus there exists a set of eight (ω, k) solutions, modes of oscillation of the field V that have the same energy in the moving frame. Note that we consider only positive comoving frequencies ω' low enough for the contour line (3.73) not to intersect with the top dispersion branch. On either side of the RIF, we either find eight propagating modes or six propagating modes and two exponentially growing and decaying modes, respectively, that take on complex ω and k.

3.2.3.3 Subluminal Intervals

We now study the nature and configuration of modes as a function of the RIF height δn and for all comoving frequencies ω'. Emission spectra with eight propagating modes on only one side of the boundary were calculated in [47, 51], where the velocity of the RIF was finely tuned to maintain such a mode structure when the RIF height was varied. In [53] we addressed the experimentally relevant case allowing for eight modes to propagate on either side of small refractive-index changes. Here we present further results for all configurations and step heights. Following on the above analyses, we focus our attention on the configurations of modes belonging to the "middle"-frequency branch in our model (3.59) (where $\lambda_2 < \lambda < \lambda_3$)—in the two materials studied later in this Thesis, this branch corresponds to the optical frequency interval. We shall henceforth refer to it as the *optical* branch, whilst the lower and higher frequency branches will be referred to as *IR* and *UV* branch, respectively. Indeed, for our set of material parameters, we find that the mode configuration only varies over the optical branch, whilst the nature of modes belonging to other branches never changes.

In particular, consider the positive frequency optical branch in the moving frame, depicted in Fig. 3.5. The black (orange) curve is the branch on the right (left) side of the RIF. The number of mode solutions depends on ω'. On either side of the RIF, there is at least one propagating optical mode for all ω'. There is also a frequency interval over which a line of constant ω' (that would be horizontal in Fig. 3.5) intersects three times with the optical branch—there, three propagating modes exist: between the two horizontal dashed black lines and two horizontal dashed orange lines in Fig. 3.5, respectively. Hereafter, these frequency intervals on either side of the RIF are referred to as the *subluminal intervals* (SLIs) $\left[\omega'_{minL}, \omega'_{maxL}\right]$ and $\left[\omega'_{minR}, \omega'_{maxR}\right]$. On all other branches, of positive or negative frequency, there always exist only one mode—i.e., one oscillatory solution to the equation of motion.

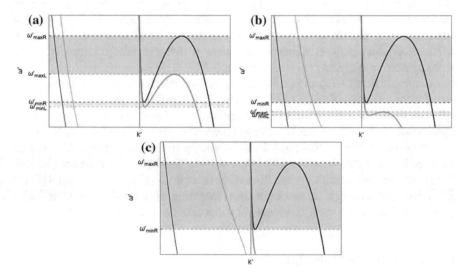

Fig. 3.5 Sellmeier dispersion relation of fused silica in a frame moving at a velocity $u = 0.66c$. Part of the optical branch is shown: branches with positive (negative) laboratory frequencies are represented by thick (thin) curves. A curve for zero refractive-index change δn is shown in black, and that for a large change, $\delta n = 0.12$ in (**a**), medium change $\delta n = 0.048$ in (**b**), small change $\delta n = 0.02$ in (**c**), is in orange. Frequency intervals corresponding to black- and white-hole analogue horizons are shaded in orange ($\left[\omega'_{maxL}, \omega'_{maxR}\right]$), and blue ($\left[\omega'_{minL}, \omega'_{minR}\right]$), respectively

3.2.3.4 Mode Configurations

Since only the optical modes change in nature (complex or oscillatory) as a function of the comoving frequency ω', studying their configurations allows to fully characterise the system. Indeed, as will be shown now, the number of oscillatory optical modes, as well as their direction of propagation with respect to the RIF, on either side of the RIF determines the essence of the boundary. That is, whether it acts as an analogue horizon for modes of the field of a given frequency.

Inside a SLI, one of the three mode solutions has a positive comoving group velocity $\frac{\partial \omega'}{\partial k'}$. This unique optical mode allows light on the right of the RIF to propagate away from it. This middle optical mode (see Fig. 3.3) on the right is called moR in what follows. The other two modes have negative comoving group velocity; they move into the boundary from the right. There is a lower (upper) optical mode denoted loR (uoR). On either side we can order the modes by the comoving wave number k' and obtain $k_{\omega'}^{loR/L} < k_{\omega'}^{moR/L} < k_{\omega'}^{uoR/L}$ (see Fig. 3.5). In the laboratory frame, this translates into $\omega_{\omega'}^{loR/L} < \omega_{\omega'}^{moR/L} < \omega_{\omega'}^{uoR/L}$. Remark that both moR and uoR have positive laboratory frame group velocity at all frequency—to an observer in the rest frame of the medium they propagate in the same direction as the RIF—whilst loR has positive laboratory frame group velocity for low ω' and negative laboratory frame group velocity for high comoving frequency. Note that, except on the positive laboratory optical frequency branch, all modes always have negative comoving group velocity.

Beyond the SLI—i.e., $\omega' \notin \left[\omega'_{min}, \omega'_{max}\right]$—only one propagating mode remains. Two complex-wave-number roots of (3.59) and (3.73) emerge as pairs of exponentially growing and decaying modes that do not propagate. For $\omega' < \omega'_{min}$ only mode uoR/L remains a propagating mode, whereas for $\omega' > \omega'_{max}$, only loR/L remains. As stated earlier, and as can also be seen in Fig. 3.5, for all comoving frequencies there is one propagating mode that belongs to the negative optical-frequency branch. This mode has a negative norm (3.63) (see Sect. 3.2.2.3) and will hereafter be referred to as noR/L.

For all magnitudes in the refractive index change δn, the subluminal intervals on either side of the RIF do not fully overlap: the SLI of the left region is, in general, different from that of the right region. For small[24] refractive index changes, the left and right SLIs overlap and there therefore exist five different combinations of modes across the RIF, also shown in Fig. 3.6: in growing order of comoving frequency, we have—

1. $\omega' < \omega'_{minL}$. One optical propagating mode (uoL/R) exists, and has negative group velocity in the moving frame, on either side of the boundary.
2. $\omega'_{minL} < \omega' < \omega'_{minR}$. On the left of the boundary, there exist three optical propagating modes (loL, moL, and uoL) whilst only mode uoR exists on the right.

---

[24]The magnitude of the refractive index change giving rise to the various mode configurations depends on the medium properties. For the sake of the argument presented in this section it suffices to identify three categories of refractive index change: small, medium, and large—exact numbers will be provided by the numerical analysis carried in Sect. (4.3).

All modes in the inhomogeneous medium have negative comoving group velocity, except for *moL* that has positive comoving group velocity on the left.

3. $\omega'_{minR} < \omega' < \omega'_{maxL}$. Three propagating modes (*loL/R, moL/R,* and *uoL/R*) exist on either side of the boundary. Mode *moL(R)* has positive comoving group velocity on the left (right) of the RIF, and all other modes have negative comoving group velocity.

4. $\omega'_{maxL} < \omega' < \omega'_{maxR}$. Only one mode, with negative comoving group velocity, exists on the left of the boundary, but modes *loL/R, moL/R,* and *uoL/R* exist on the right—with negative, positive, and negative comoving group velocity, respectively.

5. $\omega' > \omega'_{maxR}$. One propagating mode (*loL/R*) exists on either side of the boundary. All propagating modes exhibit negative group velocities.

For medium refractive index change, the SLIs on either side of the RIF do not overlap at all. There exist five different combinations of modes across the RIF, also shown in Fig. 3.6:

1. $\omega' < \omega'_{minL}$. One optical propagating mode (*uoL/R*) exists, and has negative group velocity in the moving frame, on either side of the boundary.

2. $\omega'_{minL} < \omega' < \omega'_{maxL}$. On the left of the boundary, there exist three optical propagating modes (*loL, moL,* and *uoL*) whilst only mode *uoR* exists on the right. All modes in the inhomogeneous medium have negative comoving group velocity, except for *moL* that has positive comoving group velocity on the left.

3. $\omega'_{maxL} < \omega' < \omega'_{minR}$. Only mode *loL* propagates on the left of the boundary, whilst only mode *uoR* exists on the right—all modes in the inhomogeneous medium have negative comoving group velocity.

4. $\omega'_{minR} < \omega' < \omega'_{maxR}$. Only mode *loL*, that has negative comoving group velocity, exists on the left of the boundary, but modes *loL/R, moL/R,* and *uoL/R* exist on the right—with negative, positive, and negative comoving group velocity, respectively.

5. $\omega' > \omega'_{maxR}$. One propagating mode (*loL/R*) exists on either side of the boundary. All propagating modes exhibit negative group velocities.

Finally, for a large RIF height, only three mode configurations exist. Indeed, the refractive index change is then so high that the positive frequency optical laboratory branch on the right of the boundary exhibits no pole in the moving frame (see Fig. 3.5)—no SLI exists on the left of the RIF. We then find (as was studied in [47, 50]) the following configurations:

1. $\omega' < \omega'_{minR}$. One optical propagating mode (*loL/uoR*) exists, and has negative group velocity in the moving frame, on either side of the boundary.

2. $\omega'_{minR} < \omega' < \omega'_{maxR}$. Only mode *loL*, that has negative comoving group velocity, exists on the left of the boundary, but modes *loL/R, moL/R,* and *uoL/R* exist on the right—with negative, positive, and negative comoving group velocity, respectively.

3. $\omega' > \omega'_{maxR}$. One propagating mode (*loL/R*) exists on either side of the boundary. All propagating modes exhibit negative group velocities.

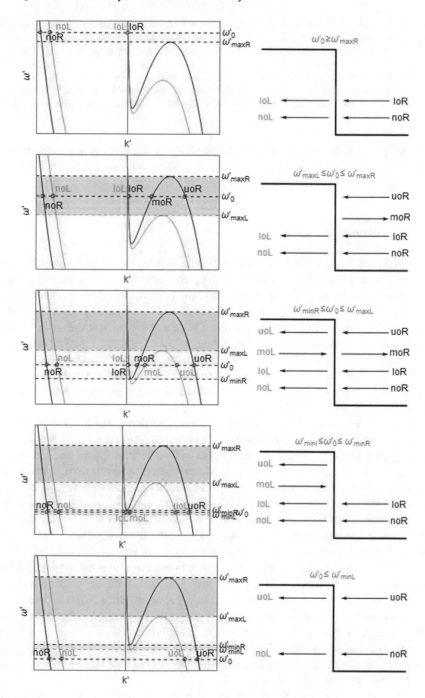

Fig. 3.6 (Continued)

◄**Fig. 3.6** (Continued) Diagrammatic explanation of the possible mode configurations for positive-
and negative-norm optical modes for various comoving frequencies in the regime of low refractive
index change. Modes are schematically sketched at the step for comoving frequency ω_0' (blue dashed
line in the dispersion diagrams). The arrows indicate the comoving group velocity of each mode.
Modes noL and noR are the only negative-norm optical modes, on the left and right of the step,
respectively. All other optical modes have a positive norm. The step acts as a black-hole-like horizon
over the orange-shaded interval, and as a white-hole-like horizon over the blue-shaded interval

In [53], we introduced and studied the physics of low refractive index changes—
that describe a typical experiment optical analogue experiment. Likewise, the study
of a medium change in the refractive index is new.

3.2.3.5 Analogy to Gravity

In Sect. 2.3 of this dissertation, we laid out the first argument of the present Thesis: a Refractive Index Front (RIF) in a dispersive medium can act as an analogue
event horizon to modes of the field. We will now build on this finding and use kinematic arguments (after the suggestion [50]) to identify the mode configurations that
reproduce the physics of curved spacetimes. We thus look back at the mode configurations identified in the Sect. 3.2.3.4 and begin with the case of the low refractive
index change. Our discussion will take place in the comoving frame, where the RIF
is stationary. Thus we omit the notation "comoving" where this leaves no doubt—for
example the direction of propagation of a mode is always considered with respect to
the stationary boundary at $\zeta = 0$.

In configurations 1 and 5, the increase in the refractive index in the right region
does not modify either the nature nor the direction of propagation of the sole optical frequency mode that exists on either side of the RIF: no optical horizon exists.
Configuration 2 is more interesting, and its description is novel: light in mode *loL*
propagates from the left into the boundary, but cannot enter the right region, because
all modes there have negative group velocity. Over the $\left[\omega_{minL}', \omega_{minR}'\right]$ frequency
interval, the boundary acts as a white hole to modes of the field as light can approach
but not enter the right region. Symmetrically, over the frequency interval of configuration 4, light experiences a black-hole horizon at the RIF as it cannot propagate
to the right from beyond the RIF. Finally, configuration 3 is similar to 1 and 5 in
that the step in the refractive index does not affect either the nature or the direction of propagation of the optical modes. It is, however, slightly different from them
in that modes with negative and positive group velocity exist on either side of the
RIF: although the RIF is not a one-way door (as in configurations 2 and 4) and thus
no horizon exists for waves of this frequency, the situation is somewhat analogous
to gravitational disturbances such as gravitational waves. The latter comment is an
original observation of this Thesis.

In the case of a large refractive index change, only configurations 1 (with *loL*
instead of *uoL*), 4 and 5 remain. For medium δn magnitude, all mode configurations
are identical to the low δn case, except for configuration three where only one mode

can propagate on either side of the RIF, with *loL* on the left and *uoR* on the right. There, the mismatch created by the increase in the refractive index renders the system horizonless.

3.2.3.6   Optical Horizons

In configuration 2 (4), the region on the right (left) of the boundary corresponds to the inner region, whilst that on the left (right) corresponds to the outer region, respectively, of the analogue horizon. Consider configuration 4: on the left of the boundary, light can only propagate to the left—in only one direction, in analogy with the interior region of a black hole described by the Painlevé-Gullstrand metric (2.38) where the spacetime flow is superluminal. In contrast, on the right of the boundary, light can propagate in both directions (in analogy with a subluminal flow of spacetime). The symmetrical analogy holds for configuration 2. This analogy to black- or white-hole physics stems from the disturbance in the refractive index, which plays the same role as the geometrical disturbance in the vicinity of a black hole.

So, according to our intuition of Sect. 2.3, light in a dispersive medium can be made to interact with itself so as to create analogue horizons. Note that we also discovered that a RIF acts simultaneously as a black hole, white hole, and no horizon boundary (although over different discrete frequency intervals). In the next section of this dissertation, we will proceed to quantising the field theory, by resorting to the tools of quantum field theory in curved spacetime presented in Sect. 3.1. We will thus quantise for small perturbations (the plane wave modes of the inhomogeneous medium) on a classical geometrical background (the refractive index increase at the RIF). This will reveal how, in total analogy with black hole physics, fluctuations of the quantum vacuum at the RIF give rise to spontaneous emission of light.

3.2.4   Scattering of the Quantum Vacuum at the RIF

In the previous section, we have derived solutions on either side of the RIF, we now construct "global" solutions, i.e., solutions to the equation of motion that are valid in both regions. These modes correspond to waves scattering at the RIF, and they describe the conversion of an incoming field, even in the quantum vacuum state, to scattered fields in both regions. We follow the canonical approach introduced in [56], developed in the 1990s in [67–72], and used in [47] and [48, 49] to construct these modes and their scattering matrix and then to quantise the solutions to find photon fluxes due to spontaneous particle creation.

3.2.4.1   Mode Matching Across the Boundary

We now proceed to match the asymptotic stationary modes (3.57) across the refractive index boundary at $\zeta = 0$. Since they exist in only one of the two homogeneous regions separated by the RIF, these modes will henceforth be denoted local modes (LMs).

On physical grounds, we consider all fields, conjugate momenta and time derivatives to be finite. By construction of the model, the elastic constant and inertia of the polarisation fields are, respectively, discontinuous and continuous at the interface between the two homogeneous media. In the near-interface region, we gain insight in the behaviour of the fields and conjugate momenta by integrating the equations of motion (3.56) over time. We begin with the third equation of (3.58): we integrate with respect to the spatial coordinate about $\zeta = 0$ from $-\epsilon$ to $+\epsilon$, taking the limit $\epsilon \to 0$,

$$\int_{-\epsilon}^{+\epsilon} i\omega' \tilde{\Pi}_A d\zeta = \int_{-\epsilon}^{+\epsilon} \frac{\tilde{A}''}{4\pi} d\zeta + \int_{-\epsilon}^{+\epsilon} \sum_{i=1}^{3} \frac{\kappa_i (2\pi c)^2}{\gamma c \lambda_i^2} \left(\tilde{\Pi}_{P_i} - \gamma \frac{\tilde{A}}{c} \right) d\zeta. \quad (3.74)$$

All finite terms integrate to zero for the limit $\epsilon \to 0$, thus

$$\int_{-\epsilon}^{+\epsilon} \sum_{i=1}^{3} \frac{\kappa_i (2\pi c)^2}{\gamma c \lambda_i^2} \left(\tilde{\Pi}_{P_i} - \gamma \frac{\tilde{A}}{c} \right) d\zeta \to 0, \quad (3.75)$$

and (3.74) yields

$$\int_{-\epsilon}^{+\epsilon} \frac{\tilde{A}''}{4\pi} d\zeta = 0, \quad (3.76)$$

A'' is finite. Thus the vector potential A is continuously differentiable: $A_L = A_R$ and $A'_L = A'_R$. Proceeding similarly with the second equation of (3.58) leads to the condition

$$u \int_{-\epsilon}^{+\epsilon} \tilde{P}'_i d\zeta = 0 \quad (3.77)$$
$$\Rightarrow P_{iL} = P_{iR}.$$

That is, the polarisation fields are continuous across the interface. We apply the same process to the fourth equation of (3.58): all the terms being finite, integrating and subsequently taking the limit $\epsilon \to 0$ shows that the Π_{P_i}s are continuous as well. Glancing again at the second equation of (3.58) and noticing that all the terms except P'_i are continuous we realise that the spatial derivatives of the polarisation fields are also continuous: $P'_{iL} = P'_{iR}$. Finally, turning back to the fourth equation of (3.58), in which both P_i and Π_{P_i} ($k'\Pi_{P_i}$) are continuous, we see that the discontinuity in κ_i implies that the term $\partial_\zeta(u\Pi_{P_i})$ must carry a discontinuity. Equating the Hamilton equations for each side of the step by identifying $\dot{\Pi}_{P_iL} = \dot{\Pi}_{P_iR}$ yields

$$\left(\dot{\Pi}_{P_i L} - u \Pi'_{P_i L}\right) = \frac{-P_{iL}}{\kappa_{iL}},$$

$$\left(\dot{\Pi}_{P_i R} - u \Pi'_{P_i R}\right) = \frac{-P_{iR}}{\kappa_{iR}}, \tag{3.78}$$

$$\Rightarrow \Pi'_{P_i R} - \Pi'_{P_i L} = \frac{P_i}{u} \left(\frac{1}{\kappa_{iR}} - \frac{1}{\kappa_{iL}}\right).$$

That is, Π'_{P_i} is discontinuous.

To sum up, we have found that the fields and their conjugate momenta are continuous at the boundary. The spatial derivatives of all fields and conjugate momenta are also continuous, with the exception of Π'_{P_i}. Furthermore, looking at the Hamilton equations of motion (3.58), we see that the finiteness of the temporal derivatives of the fields imply that they are continuous.

3.2.4.2 Global Modes of the Inhomogeneous Medium

We now use the S-matrix formalism to relate incoming and outgoing fields at the RIF. We thus seek bases of *in* and *out* modes that live in the two regions of the inhomogeneous medium and are related by the scattering matrix. These are called global modes (GMs) We construct the GMs $\mathcal{V}$ as

$$\mathcal{V} = \sum_\alpha L^\alpha V_L^\alpha \theta(-\zeta) + \sum_\alpha R^\alpha V_R^\alpha \theta(\zeta), \tag{3.79}$$

where L^α (R^α) describes the strength of mode α on the left (right) side of the RIF. Half of the coefficients in (3.79) are constrained by the matching conditions. We consider GMs whose asymptotic decomposition comprises only a single LM with comoving frame group velocity towards (*in*) or away from (*out*) the RIF [73]. Thus there are as many of these GMs as there are propagating local modes. Half of the GMs emerge from a defining LM α that moves towards the RIF, forming global *in* modes $\mathcal{V}^{in\alpha}$. The other GMs are global *out* modes $\mathcal{V}^{out\alpha}$ if α is a LM now moving away from the RIF. The LMs are the complete physical (i.e., nondivergent) solutions in the asymptotic regions, thus the sets of $\mathcal{V}^{in}$ and $\mathcal{V}^{out}$ modes are two basis sets of modes. Hence the scattering matrix S is the transformation of modes from the out basis to the in basis:

$$\mathcal{V}^{out\alpha} = \sum_\beta S_{\alpha\beta} \mathcal{V}^{in\beta}. \tag{3.80}$$

Scattering and spontaneous photon creation occur as the input vacuum state does not correspond to the vacuum state in the out basis (see Sect. 3.1.2.3); that is, the spontaneous emission follows from S, that governs all mode conversion.

3.2.4.3   Quantum Field Theory

We postulate the equivalent of the standard equal-time commutation relations on the fields A and P_i and thus quantise the local field modes and their momenta:

$$[A(\zeta), \Pi_A(\zeta')] = i\hbar\delta(\zeta - \zeta'), \tag{3.81}$$

$$[P_i(\zeta), \Pi_{P_j}(\zeta')] = i\hbar\delta_{ij}\delta(\zeta - \zeta'). \tag{3.82}$$

We expand the field V on the basis of local frequency eigenmodes

$$V = \int d\omega' \sum_{\alpha} \left(V^{\alpha}_{\omega'} \hat{a}^{\alpha}_{\omega'} + V^{\alpha*}_{\omega'} \hat{a}^{\alpha\dagger}_{\omega'} \right) \tag{3.83}$$

that are properly normalised with respect to the scalar product (3.64) under the condition [47]

$$\left| \left\langle V^{\alpha_1}_{\omega'_1}, V^{\alpha_2}_{\omega'_2} \right\rangle \right| = \delta(\omega'_2 - \omega'_1)\delta_{\alpha_2\alpha_1}. \tag{3.84}$$

According to our quantum theory for the field (see Sect. 3.1.2.3), the operators $\hat{a}^{\alpha}_{\omega'}$ and $\hat{a}^{\alpha\dagger}_{\omega'}$ are the annihilation and creation operators of the field mode α.

Alternatively, we can expand the field over positive frequencies only, including negative-norm modes in the expansion:

$$V = \int_0^{\infty} d\omega' \left(\sum_{\alpha \in P} V^{\alpha}_{\omega'} \hat{a}^{\alpha}_{\omega'} + \sum_{\alpha \in N} V^{\alpha}_{\omega'} \hat{a}^{\alpha\dagger}_{\omega'} \right) + H.c., \tag{3.85}$$

where $P(N)$ is the set of modes of positive (negative) norm. We quantise the GMs by writing the global field $\mathcal{V}$ in the basis of global *in* modes:

$$\mathcal{V} = \int_0^{\infty} d\omega' \left(\sum_{\alpha \in P} V^{in\alpha}_{\omega'} \hat{a}^{in\alpha}_{\omega'} + \sum_{\alpha \in N} V^{in\alpha}_{\omega'} \hat{a}^{in\alpha\dagger}_{\omega'} \right) + H.c., \tag{3.86}$$

or global *out* modes:

$$\mathcal{V} = \int_0^{\infty} d\omega' \left(\sum_{\alpha \in P} V^{out\alpha}_{\omega'} \hat{a}^{out\alpha}_{\omega'} + \sum_{\alpha \in N} V^{out\alpha}_{\omega'} \hat{a}^{out\alpha\dagger}_{\omega'} \right) + H.c., \tag{3.87}$$

The expansion (3.86) for *in* and (3.87) for *out* modes defines the annihilation and creation operators for the global modes, as well as the transformation between *in* and *out* creation and annihilation operators of the field. Let $\hat{A}^{in}$ be the row vector containing all the annihilation and creation operators for positive- and negative-norm global *in* modes, respectively, and $\hat{A}^{out}$ be the corresponding variable for the *out* modes, then the transformation of operators follows from the definition of S [53]:

$$\hat{A}^{out} = S\hat{A}^{in}. \tag{3.88}$$

3.2.4.4 Scattering of Vacuum States

Having quantised the sets of global *in* and *out* modes, we can use scattering theory to calculate the expectation value in *out* modes of positive or negative norm when *in* modes are in the vacuum state.

Denoting α $(\bar{\alpha})$ as a mode of same (opposite) sign in norm as α_1, the incoming state is defined as

$$|0_{in}\rangle = \hat{a}_{\omega'}^{in\alpha}|0_\alpha\rangle \otimes \hat{a}_{\omega'}^{in\bar{\alpha}}|0_{\bar{\alpha}}\rangle = 0. \tag{3.89}$$

This state is in the vacuum state defined by the destruction operators associated with the *in* modes of positive and negative norm. The number of particles operator in an *out* mode α_1 is $\hat{N}^{\alpha_1} = \hat{a}^{out\alpha_1\dagger}\hat{a}^{out\alpha_1}$. It can be written out by identifying the annihilation and creation operators of the *out* mode from Eq. (3.87):

$$
\begin{aligned}
\hat{N}^{\alpha_1} &= \left(\sum_\alpha \beta^{\alpha\alpha_1\star}\hat{a}^{\alpha\dagger} + \sum_{\bar{\alpha}} \beta^{\bar{\alpha}\alpha_1}\hat{a}^{\bar{\alpha}}\right)\left(\sum_\alpha \beta^{\alpha\alpha_1}\hat{a}^\alpha + \sum_{\bar{\alpha}} \beta^{\bar{\alpha}\alpha_1\star}\hat{a}^{\bar{\alpha}\dagger}\right) \\
&= \sum_{\alpha\alpha'} \beta^{\alpha\alpha_1\star}\beta^{\alpha'\alpha_1}\hat{a}^{\alpha\dagger}\hat{a}^{\alpha'} + \sum_{\bar{\alpha}\bar{\alpha}'} \beta^{\bar{\alpha}\alpha_1}\beta^{\bar{\alpha}'\alpha_1\star}\hat{a}^{\bar{\alpha}}\hat{a}^{\bar{\alpha}'\dagger} + \\
&\quad \sum_{\alpha\bar{\alpha}'} \beta^{\alpha\alpha_1\star}\beta^{\bar{\alpha}'\alpha_1\star}\hat{a}^{\alpha\dagger}\hat{a}^{\bar{\alpha}'\dagger} + \sum_{\bar{\alpha}\alpha'} \beta^{\bar{\alpha}\alpha_1}\beta^{\alpha'\alpha_1}\hat{a}^{\bar{\alpha}}\hat{a}^{\alpha'}.
\end{aligned} \tag{3.90}
$$

Whence the expectation value for the number of photons in an *out* mode is $\left\langle\hat{N}^{\alpha_1}\right\rangle = \left\langle 0_{in}\left|\hat{N}^{\alpha_1}\right|0_{in}\right\rangle$. We begin with the second term (all the mixed terms go to zero):

$$
\begin{aligned}
\langle 0_{in}|\sum_{\bar{\alpha}\bar{\alpha}'} \beta^{\bar{\alpha}\alpha_1}\beta^{\bar{\alpha}'\alpha_1\star}\hat{a}^{\bar{\alpha}}\hat{a}^{\bar{\alpha}'\dagger}|0_{in}\rangle &= \sum_{\bar{\alpha}\bar{\alpha}'} \beta^{\bar{\alpha}\alpha_1}\beta^{\bar{\alpha}'\alpha_1\star} \times \\
&\quad \left(\langle 0_{in}|\hat{a}^{\bar{\alpha}\dagger}\hat{a}^{\bar{\alpha}'}|0_{in}\rangle + \langle 1|1\rangle_{in}\,\delta_{\bar{\alpha}\bar{\alpha}'}\right) \\
&= \sum_{\bar{\alpha}} |\beta^{\bar{\alpha}\alpha_1}|^2,
\end{aligned} \tag{3.91}
$$

because the $\langle 1|1\rangle_{in}$ term for any mode is nothing but $\langle 0_{in}|\hat{a}\hat{a}^\dagger|0_{in}\rangle$ and for the same mode, $\hat{a}\hat{a}^\dagger - \hat{a}^\dagger\hat{a} = \delta(0)$, thus

$$\langle 1|1\rangle_{in} = \langle 0_{in}|\hat{a}^\dagger\hat{a}|0_{in}\rangle + \langle 0_{in}|\delta(0)|0_{in}\rangle = \delta(0). \tag{3.92}$$

Furthermore, by (3.89), the first term in the parentheses of (3.91) is zero. Likewise, all the other terms of (3.90) are zero[25]—therefore, $\langle \hat{N}^{\alpha_1} \rangle = \sum_{\bar{\alpha}} |\beta^{\bar{\alpha}\alpha_1}|^2$. We obtain the flux density of photons $I_{\omega'}^{\prime\alpha}$ in mode α, the number of particles per unit time $\Delta\tau$ and bandwidth in the moving frame,

$$I_{\omega'}^{\prime\alpha} = \frac{\langle \hat{N}^{\alpha} \rangle}{\Delta\tau} = \frac{1}{\Delta\tau} \sum_{\bar{\alpha}} |S^{\bar{\alpha}\alpha}|^2 \tag{3.93}$$

Note that this result is different from that obtained by Finazzi and Carusotto in [47]: they had an unargumented factor 2π in their single-mode calculation and their result was less general—we present here the correct, general, result by means of a detailed algebraic calculation that cannot be found elsewhere.

3.2.5 Conclusion and Discussion

Before progressing to the numerical computation of the flux (3.93), it is worth commenting on some aspects of the physics of optical event horizon. In deriving (3.93), we found that, as a result of the mixing of positive and negative norm modes of the field at the RIF, light would be spontaneously emitted from the vacuum. This effect is ruled by the scattering matrix S that relates *in* to *out* modes. In the scheme of optical analogues, the event at which light is emitted is very well located in space: light is emitted at the RIF (in the case studied here, at the interface between the two regions of homogeneous refractive index). This is in contrast with the astrophysical case for which the *exact* event at which Hawking radiation is emitted cannot be easily (or at all) established—see Sect. 3.1.2.3 for a discussion of this issue. Interestingly, this is not the only advantage of the optical scheme: the frequency of the *in* and *out* modes is ruled by the dispersion relation of the medium, and both sets of modes feature only finite frequencies (with the exception of the diverging modes). Thus, dispersion limits the effect of frequency shifting of the potential on the modes (the increase in the refractive index that effectively is the curvature of spacetime for modes of the inhomogeneous medium)—dispersion seems to be the analogue phenomenon to TransPlanckian physics but here the effect is fully understood. Moreover, in the present case, the derivation of the *out* flux density (3.93) clarifies greatly the phenomenon of spontaneous emission of light from the vacuum: it results from the mixing of modes of positive and negative norm at the RIF and yields (quasi-pairwise) emission into modes of positive and negative norm. The study of the optical analogue thus enables us to cast light on various aspects of spontaneous emission from the vacuum at the horizon, and to better understand the mechanism of Hawking radiation. To this end, the next Chapter will present the algorithm that we created to implement

---

[25]The commutation of the *out* modes on the *in* modes gives zero and all the mixed terms go to zero.

(3.93) and calculate spectra of emission for any frequency, in all modes, for all mode configurations, and for a variety of refractive index changes δn in both the moving and laboratory frame.

References

1. B. Carter, Republication of: black hole equilibrium states: part i analytic and geometric properties of the kerr solutions. Gen. Relat. Gravit. **41**(12), 2873–2938 (2009)
2. P.C.W. Davies, Thermodynamics of black holes. Rep. Prog. Phys. **41** (1978)
3. N.D. Birrell, P.C.W. Davies, Quantum fields in curved space, repr edn. Cambridge monographs on mathematical physics. (Cambridge Univ. Press, Cambridge, 1994)
4. T. Jacobson, *Introductory Lectures on Black Hole Thermodynamics* (1996)
5. D. Lynden-Bell, R. Wood, A. Royal, The gravo-thermal catastrophe in isothermal spheres and the onset of red-giant structure for stellar systems. Mon. Not. R. Astron. Soc. **138**(4), 495–525 (1968)
6. P.C.W. Davies. *The Physics of Time Asymmetry*. (University of California Press, Berkeley and Los Angeles, 1977). OCLC: 232966619
7. S.W. Hawking, G.F.R. Ellis, *The Large Scale Structure of Space-time*. Cambridge monographs on mathematical physics. (Cambridge Univ. Press, Cambridge, 21. printing edition, 2008). OCLC: 552219048
8. S.W. Hawking, Black holes in general relativity. Commun. Math. Phys. **25**(2), 152–166 (1972)
9. S. Chandrasekhar, The maximum mass of ideal white dwarfs. Astrophys. J. **74**, 81 (1931)
10. R. Penrose, Gravitational collapse: the role of general relativity. Riv. Nuovo Cim. **1**, 252–276 (1969)
11. G.W. Gibbons, Vacuum polarization and the spontaneous loss of charge by black holes. Commun. Math. Phys. **44**(3), 245–264 (1975)
12. R. Penrose, G. Collapse, S.-T. Singularities, Phys. Rev. Lett. **14**(3), 57–59 (1965)
13. J.M. Bardeen, B. Carter, S.W. Hawking, The four laws of black hole mechanics. Commun. Math. Phys. **31**(2), 161–170 (1973)
14. H.B. Callen, *Thermodynamics and an Introduction to Thermostatistics*, 2nd edn. (Wiley, New York, 1985)
15. J.D. Bekenstein, Black holes and entropy. Phys. Rev. D **7**(8), 2333–2346 (1973)
16. L. Smarr, Mass formula for kerr black holes. Phys. Rev. Lett. **30**, 71–73 (1973)
17. S.W. Hawking, Black holes and thermodynamics. Phys. Rev. D **13**(2), 191–197 (1976)
18. R. Penrose, Black holes and gravitational theory. Nature **236**(5347), 377–380 (1972)
19. Y.B. Zel'dovich, Pis'ma. Zh. Eksp. Teor. Fiz **12**, 443 (1970)
20. C.W. Misner, Interpretation of gravitational-wave observations. Phys. Rev. Lett. **28**(15), 994–997 (1972)
21. Y.B. Zel'dovich, Amplification of cylindrical electromagnetic waves reflected from a rotating body. Sov. Phys. JETP **35**(6) (1972)
22. A.A. Starobinski, Y.B. Zel'dovich, Pis'ma. Zh. Eksp. Teor. Fiz **26**, 373 (1977)
23. W.G. Unruh, Origin of the particles in black-hole evaporation. Phys. Rev. D **15**(2), 365–369 (1977)
24. K.S. Thorne, *Black Holes and Time Warps: Einstein's Outrageous Legacy*. (1994). OCLC: 28147932
25. S.W. Hawking, Black hole explosions? Nature **248**(5443), 30–31 (1974)
26. S.W. Hawking, Particle creation by black holes. Commun. Math. Phys. **43**(3), 199–220 (1975)
27. C.W. Misner, K.S. Thorne, J.A. Wheeler, *Gravitation*. (W.H. Freeman, San Francisco, 1973)
28. L. Parker, Probability distribution of particles created by a black hole. Phys. Rev. D **12**(6), 1519–1525 (1975)

29. L. Parker, Cincinnati, Proceedings, Asymptotic Structure of Space-time (1976), pp. 107–226
30. N.N. Bogoljubov, On a new method in the theory of superconductivity. J. Exp. Theor. Phys. **34**(1) (1958)
31. S. Corley, T. Jacobson, Hawking spectrum and high frequency dispersion. Phys. Rev. D **54**(2), 1568–1586 (1996)
32. T. Jacobson, Black-hole evaporation and ultrashort distances. Phys. Rev. D **44**(6), 1731–1739 (1991)
33. P.C.W. Davies, S.A. Fulling, W.G. Unruh, Energy-momentum tensor near an evaporating black hole. Phys. Rev. D **13**(10), 2720–2723 (1976)
34. T.G. Philbin, C. Kuklewicz, S. Robertson, S. Hill, F. König, U. Leonhardt, Fiber-optical analog of the event horizon. Science **319**(5868), 1367–1370 (2008)
35. W.G. Unruh, Experimental black-hole evaporation? Phys. Rev. Lett. **46**(21), 1351–1353 (1981)
36. A.A. Penzias, R.W. Wilson, A measurement of excess antenna temperature at 4080 Mc/s. Astrophys. J. **142**, 419 (1965)
37. F. Belgiorno, S.L. Cacciatori, M. Clerici, V. Gorini, G. Ortenzi, L. Rizzi, E. Rubino, V.G. Sala, D. Faccio, Hawking radiation from ultrashort laser pulse filaments. Phys. Rev. Lett. **105**(20) (2010)
38. D. Bermudez, U. Leonhardt, Hawking spectrum for a fiber-optical analog of the event horizon. Phys. Rev. A **93**(5) (2016)
39. K.E. Webb, M. Erkintalo, Y. Xu, N.G.R. Broderick, J.M. Dudley, G. Genty, S.G. Murdoch, Nonlinear optics of fibre event horizons. Nat. Commun. **5**, 4969 (2014)
40. E. Rubino, F. Belgiorno, S.L. Cacciatori, M. Clerici, V. Gorini, G. Ortenzi, L. Rizzi, V.G. Sala, M. Kolesik, D. Faccio, Experimental evidence of analogue hawking radiation from ultrashort laser pulse filaments. New J. Phys. **13**(8), 085005 (2011)
41. E. Rubino, A. Lotti, F. Belgiorno, S.L. Cacciatori, A. Couairon, U. Leonhardt, D. Faccio, Soliton-induced relativistic-scattering and amplification. Sci. Rep. **2** (2012)
42. M. Petev, N. Westerberg, D. Moss, E. Rubino, C. Rimoldi, S.L. Cacciatori, F. Belgiorno, D. Faccio, Blackbody emission from light interacting with an effective moving dispersive medium. Phys. Rev. Lett. **111**(4) (2013)
43. S. Liberati, A. Prain, M. Visser, Quantum vacuum radiation in optical glass. Phys. Rev. D **85**(8) (2012)
44. F. Belgiorno, S.L. Cacciatori, F.D. Piazza, Perturbative photon production in a dispersive medium. Eur. Phys. J. D **68**(5) (2014)
45. S.F. Wang, A. Mussot, M. Conforti, A. Bendahmane, X.L. Zeng, A. Kudlinski, Optical event horizons from the collision of a soliton and its own dispersive wave. Phys. Rev. A **92**(2) (2015)
46. D. Faccio, S. Cacciatori, V. Gorini, V.G. Sala, A. Averchi, A. Lotti, M. Kolesik, J.V. Moloney, Analogue gravity and ultrashort laser pulse filamentation. EPL (Europhys. Lett.) **89**(3), 34004 (2010)
47. S. Finazzi, I. Carusotto, Quantum vacuum emission in a nonlinear optical medium illuminated by a strong laser pulse. Phys. Rev. A **87**(2) (2013)
48. F. Belgiorno, S.L. Cacciatori, F.D. Piazza, Hawking effect in dielectric media and the hopfield model. Phys. Rev. D **91**(12) (2015)
49. M.F. Linder, R. Schützhold, W.G. Unruh, Derivation of hawking radiation in dispersive dielectric media. Phys. Rev. D **93**(10) (2016)
50. S. Finazzi, I. Carusotto, Kinematic study of the effect of dispersion in quantum vacuum emission from strong laser pulses. Eur. Phys. J. Plus **127**(7) (2012)
51. S. Finazzi, I. Carusotto, Spontaneous quantum emission from analog white holes in a nonlinear optical medium. Phys. Rev. A **89**(5) (2014)
52. S. Robertson, Integral method for the calculation of Hawking radiation in dispersive media. II. Asymmetric asymptotics. Phys. Rev. E **90**(5) (2014)
53. M. Jacquet, F. König, Quantum vacuum emission from a refractive-index front. Phys. Rev. A **92**(2) (2015)
54. J.J. Hopfield, Theory of the contribution of excitons to the complex dielectric constant of crystals. Phys. Rev. **112**(5), 1555–1567 (1958)

55. R. Schützhold, G. Plunien, G. Soff, Dielectric black hole analogs. Phys. Rev. Lett. **88**, 061101 (2002)
56. U. Fano, Atomic theory of electromagnetic interactions in dense materials. Phys. Rev. **103**(5), 1202–1218 (1956)
57. S.I. Pekar, Theory of electromagnetic waves in a crystal with excitons. J. Phys. Chem. Solids **5**(1–2), 11–22 (1958)
58. C. Kittel, *Introduction to Solid State Physics*, 8th edn. (Wiley, Hoboken, NJ, 2005)
59. C. Cohen-Tannoudji, J. Dupont-Roc, G. Grynberg, *Photons and Atoms: Introduction to Quantum Electrodynamics*. Physics textbook. (Wiley, Weinheim, nachdr. edition, 2004). OCLC: 254806943
60. E. Noether, Invariante variationsprobleme. Nachr. von der Ges. der Wiss zu Gött. Math.-Phys. Kl. **1918**, 235–257 (1918)
61. M. Jacquet, Quantum vacuum emission at the event horizon. M.Sc. thesis, University of St Andrews, St Andrews, 2013
62. S. Weinfurtner, E.W. Tedford, M.C.J. Penrice, W.G. Unruh, G.A. Lawrence, Measurement of stimulated hawking emission in an analogue system. Phys. Rev. Lett. **106**(2) (2011)
63. G. Rousseaux, C. Mathis, P. Maïssa, T.G. Philbin, U. Leonhardt, Observation of negative-frequency waves in a water tank: a classical analogue to the hawking effect? New J. Phys. **10**(5), 053015 (2008)
64. E. Rubino, J. McLenaghan, S.C. Kehr, F. Belgiorno, D. Townsend, S. Rohr, C.E. Kuklewicz, U. Leonhardt, F. König, D. Faccio, Negative-frequency resonant radiation. Phys. Rev. Lett. **108**(25) (2012)
65. J.S. McLenaghan, Negative frequency waves in optics: control and investigation of their generation and evolution. Ph.D. thesis, University of St Andrews, St Andrews, 2014
66. J. McLenaghan, F. König, Few-cycle fiber pulse compression and evolution of negative resonant radiation. New J. Phys. **16**(6), 063017 (2014)
67. S.M. Barnett, B. Huttner, R. Loudon, Spontaneous emission in absorbing dielectric media. Phys. Rev. Lett. **68**(25), 3698–3701 (1992)
68. B. Huttner, J.J. Baumberg, S.M. Barnett, Canonical quantization of light in a linear dielectric. Europhys. Lett. (EPL) **16**(2), 177–182 (1991)
69. B. Huttner, S.M. Barnett, Dispersion and loss in a hopfield dielectric. Europhys. Lett. (EPL) **18**(6), 487–492 (1992)
70. R. Matloob, R. Loudon, S.M. Barnett, J. Jeffers, Electromagnetic field quantization in absorbing dielectrics. Phys. Rev. A **52**(6), 4823–4838 (1995)
71. S.M. Barnett, R. Matloob, R. Loudon, Quantum theory of a dielectric-vacuum interface in one dimension. J. Mod Opt. **42**(6), 1165–1169 (1995)
72. D.J. Santos, R. Loudon, Electromagnetic-field quantization in inhomogeneous and dispersive one-dimensional systems. Phys. Rev. A **52**(2), 1538–1549 (1995)
73. J. Macher, R. Parentani, Black/white hole radiation from dispersive theories. Phys. Rev. D **79**(12) (2009)

Chapter 4
Analytics and Numerics

In this chapter we calculate spectra of emission from the vacuum and study in particular conditions over which the kinematics of a moving disturbance in the refractive index of a medium are analogous to the flow of spacetime in the vicinity of a black hole. As we saw in Chaps. 2 and 3 the kinematics of waves in analogue systems is dominated by dispersion [1]. This phenomenon regularises the phase singularities at the horizon (analogue systems do not suffer from the Transplanckian Problem) but also renders the wave equations less amenable to analytical techniques.

On the other hand, numerical techniques such as finite difference time domain (FDTD) wave packet simulations [2] or Monte Carlo methods [3] can handle the complications due to dispersion and straightforwardly evolve an initial state in time. Such methods are however computationally expensive and do not yield a spectrum directly. There also exist analytical methods, that are restricted to a fixed frequency and situations in which the background varies slowly in comparison with dispersion [4, 5], or some that can only study dispersion relations that are polynomials of low degree [6, 7]. The latter provide numerical solutions of the ordinary differential equation (ODE) in position space provided that no exponentially divergent waves exist and that the gradient of background change is low. However, dispersion relations that reproduce the refractive index of materials are usually more complicated than this, and optical experiments typically rely on a large gradient in the background. In particular, when the background change becomes so steep that it can be approximated by a step-like discontinuous function, the solution can be found analytically by matching the plane wave solutions on either side of the interface [8–13].

The analytical method we present here relies on an analytical study of the plane wave solutions to a complicated dispersion relation that realistically reproduces the material properties of fused silica or optical fibres, for example (see Chap. 3), in a one-dimensional background. We study the specific case of a step-like discontinuity in the refractive index of a dispersive medium. The method directly and efficiently yields a spectrum, unlike the above-mentioned numerical techniques, and can be generalised to considerations of rapidly varying background, unlike the above-mentioned

© Springer International Publishing AG, part of Springer Nature 2018   97
M. J. Jacquet, *Negative Frequency at the Horizon*, Springer Theses,
https://doi.org/10.1007/978-3-319-91071-0_4

analytical methods. Contrarily to the direct solution of an ODE in position space, it is not restricted to a simple polynomial dispersion relation.

Ideally, one would wish to compute spectra for the optical fibre that will be used in the experiment presented in Chap. 5. Unfortunately, the dispersion relation of usable Photonic Crystal Fibres (PCF) cannot easily be cast into a Sellmeier form. This is due to the lack of theoretical knowledge of the fibres. Indeed, the manufacturer provides data for the zero dispersion wavelength of the fibres, as well as experimentally measured dispersion curves—these have then to be experimentally verified in the laboratory. The result is a discrete set of data points that describe the dispersion of the fibre, and not an analytical relation like those that the present algorithm can handle (in other words, one does not obtain the elastic constant or resonant frequency of the medium by experimental means). It is possible to fit the experimentally-acquired data with a theoretical Sellmeier dispersion but, in the case of the PCFs that could be used in the experiment, this yielded unphysical results over some frequency ranges. Therefore, the development and usage of the method presented therein will be based on a material for which the theoretical elastic and resonant frequency constants are known. We will use fused silica, as in the literature (see, for example, [14]). Incidentally, this shall allow for checking the present results against the literature.

4.1   Analytical Description of Scattering at the RIF

In the previous chapter of this dissertation, we arrived at an expression for the scattering matrix, that describes the conversion of an incoming field to an outgoing field. We now want to devise an analytical method that, from the solutions to the dispersion relation in each homogeneous media, will allow for building the global solutions used in calculating the scattering coefficients between incoming and outgoing fields at the interface.

For this purpose, we consider a single interface: a step in the refractive index separating two homogeneous regions, as schematically depicted in Fig. 3.4. As we saw in Sect. 3.2.4, at each comoving frequency ω', we find 8 mode solutions of the fields equations (3.56) on either side of the interface. In Sect. 3.2.3 we then found that, for a given height of the step (change in the refractive index) there were different, and distinct, comoving-frequency intervals in which 6 or 8 of the mode solutions in either region would be oscillatory modes of the field. When there would be only 6 oscillatory solutions, the remaining two would have complex ω' and k'—that is they would be exponentially growing or decaying waves. Thus, as a function of comoving frequency, we found 5 mode configurations, depending on the number of oscillatory solutions on either side of the interface. In what follows, we shall refer to all mode-solutions (the oscillatory and non-oscillatory solutions alike) as "modes" and only specify their nature where necessary. We called these modes "local modes" (LMs) because they exist in the homogeneous regions on either side of the boundary.

The electromagnetic and polarisation fields and their derivatives in a homogeneous regions are related by Eq. (3.58). We also established that the electromagnetic field and polarisation fields, and their first spatial derivatives, could be matched at the interface by:

$$
\begin{pmatrix} \Pi_A \\ \Pi_{P_1} \\ \Pi_{P_2} \\ \Pi_{P_3} \\ \partial_\zeta \Pi_A \\ \partial_\zeta \Pi_{P_1} \\ \partial_\zeta \Pi_{P_2} \\ \partial_\zeta \Pi_{P_3} \end{pmatrix} =
\begin{pmatrix}
-i\frac{\omega'}{4\pi c^2} & 0 & 0 & 0 & 0 & 0 & 0 & 0 \\
\frac{\gamma}{c} & -i\frac{\omega'\gamma^2}{\kappa_1\Omega_1^2} & 0 & 0 & 0 & -\frac{u\gamma^2}{\kappa_1\Omega_1^2} & 0 & 0 \\
\frac{\gamma}{c} & 0 & -i\frac{\omega'\gamma^2}{\kappa_2\Omega_2^2} & 0 & 0 & 0 & -\frac{u\gamma^2}{\kappa_2\Omega_1^2} & 0 \\
\frac{\gamma}{c} & 0 & 0 & -i\frac{\omega'\gamma^2}{\kappa_3\Omega_3^2} & 0 & 0 & 0 & -\frac{u\gamma^2}{\kappa_3\Omega_1^2} \\
0 & 0 & 0 & 0 & -i\frac{\omega'}{4\pi c^2} & 0 & 0 & 0 \\
-i\frac{\omega'\gamma}{cu} & \left(\frac{1}{\kappa_1}-\frac{\omega'^2\gamma^2}{\kappa_1\Omega_1^2}\right)\frac{1}{u} & 0 & 0 & 0 & i\frac{\omega'\gamma^2}{\kappa_1\Omega_1^2} & 0 & 0 \\
-i\frac{\omega'\gamma}{cu} & 0 & \left(\frac{1}{\kappa_2}-\frac{\omega'^2\gamma^2}{\kappa_2\Omega_2^2}\right)\frac{1}{u} & 0 & 0 & 0 & i\frac{\omega'\gamma^2}{\kappa_2\Omega_2^2} & 0 \\
-i\frac{\omega'\gamma}{cu} & 0 & 0 & \left(\frac{1}{\kappa_3}-\frac{\omega'^3\gamma^2}{\kappa_3\Omega_3^2}\right)\frac{1}{u} & 0 & 0 & 0 & i\frac{\omega'\gamma^2}{\kappa_3\Omega_3^2}
\end{pmatrix}
\begin{pmatrix} A \\ P_1 \\ P_2 \\ P_3 \\ \partial_\zeta A \\ \partial_\zeta P_1 \\ \partial_\zeta P_2 \\ \partial_\zeta P_3 \end{pmatrix} .
$$

(4.1)

Henceforth, the last vector of (4.1) (that contains the fields and their first spatial derivatives) will be called $\vec{W}$.

In what follows, we will study the relationship, defined by the scattering matrix, between the incoming and outgoing field for each of the mode configurations (found in Sect. 3.2.3), as functions of the comoving frequency. In doing so, we will detail the analytical method used in [15] to calculate the scattering matrix from the matching conditions (4.1) in all possible mode configurations. We will then return to the dispersion relation, and heuristically construct an algorithm that implements the scattering matrix, to calculate the spectrum of light spontaneously emitted from the vacuum, as it can be observed in the laboratory frame—which is the main theoretical and numerical result of this thesis.

4.1.1 Scattering Matrix

In the scattering matrix formalism, the incoming and outgoing fields at the interface are described in terms of global modes (GMs): there are global *in* and global *out* modes. The method we will develop will allow us to calculate the flux of emission into the *out* GMs. These are modes in which light propagates away from the interface, in either of the homogeneous regions.

GMs are constructed as linear combinations of LMs: an *out* GM is composed of one LM that has positive (negative) group-velocity in the high (low) refractive-index region and a collection of 8 LMs that have negative (positive) group-velocity in the high (low) refractive-index region. In the presence of non-oscillatory modes, either the first or one of the later 8 modes may be a non-oscillatory mode. Let us consider an example: over the black-hole-like interval (mode configuration 4, see Sect. 3.2.3), there is a unique *out* GM that allows for light to propagate away from the interface into the low refractive index region (on the right of the interface in Fig. 3.4),

Fig. 4.1 Mode decomposition of the global out mode *moR*. In this spacetime diagram, there is a unique mode that propagates away from the scatterer to the right (green arrow). In the past, 7 oscillatory-modes propagate toward the scatterer from the right and there is one non-oscillatory mode on the left of the scatterer

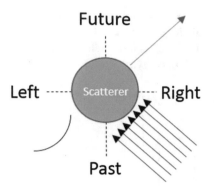

moR. Its mode decomposition is shown as the spacetime diagram in Fig. 4.1: it is a linear combination of 7 oscillatory LMs, in the right region, that have negative group-velocity, a non-oscillatory LM on the left, and a unique mode that has positive group-velocity in the right region.

The converse to the above delineation leads to constructing *in* GMs, in which light propagates toward the interface. Since there exist 8 LMs on either side of this interface, we find 8 *in* and 8 *out* GMs. These must be arranged in lowering order of laboratory-frame frequency ω to allow for a consistent treatment of the matching conditions. Given the relation between the fields, their conjugate momenta and their derivatives, see Eq. (4.1), the matching conditions are entirely determined by the fields and their derivatives only. Thus we create a matrix of the eight $\vec{W}$ LM solutions to the dispersion relation, which we call W, with

$$ W = \left(\vec{W}^{\alpha_1} \ \vec{W}^{\alpha_2} \ ... \ \vec{W}^{\alpha_8} \right), \tag{4.2} $$

with α_n, $n = 1, 2, ... 8$ the mode number, arranged in decreasing order of laboratory-frame frequency (i.e., $n = u, uo, mo, ... nu$). The $\vec{V}$ and $\vec{W}$ are related by

$$ \vec{V} = \begin{pmatrix} A \\ P_1 \\ P_2 \\ P_3 \\ \Pi_A \\ \Pi_{P_1} \\ \Pi_{P_2} \\ \Pi_{P_3} \end{pmatrix} = \begin{pmatrix} 1 & 0 & 0 & 0 & 0 & 0 & 0 & 0 \\ 0 & 1 & 0 & 0 & 0 & 0 & 0 & 0 \\ 0 & 0 & 1 & 0 & 0 & 0 & 0 & 0 \\ 0 & 0 & 0 & 1 & 0 & 0 & 0 & 0 \\ i\frac{\omega'}{4\pi c^2} & 0 & 0 & 0 & 0 & 0 & 0 & 0 \\ \frac{\gamma}{c} & -i\frac{\omega'\gamma}{\kappa_1\Omega_1^2} & 0 & 0 & 0 & -\frac{v\gamma^2}{\kappa_1\Omega_1^2} & 0 & 0 \\ \frac{\gamma}{c} & 0 & -i\frac{\omega'\gamma}{\kappa_2\Omega_2^2} & 0 & 0 & 0 & -\frac{v\gamma^2}{\kappa_2\Omega_2^2} & 0 \\ \frac{\gamma}{c} & 0 & 0 & -i\frac{\omega'\gamma}{\kappa_3\Omega_3^2} & 0 & 0 & 0 & -\frac{v\gamma^2}{\kappa_3\Omega_3^2} \end{pmatrix} \vec{W}, \tag{4.3} $$

for a field at frequency ω'. We call the matrix in (4.3) $\mathcal{U}$, and remark that $Det(\mathcal{U}) = 0$. In matrices, (4.3) reads $V = \mathcal{U}W$, Since $\vec{W}$ and $\bar{\vec{W}}$ are related by Eq. (3.57), in an identical fashion to $\vec{V}$ and $\bar{\vec{V}}$, this statement taken at $\tau = 0$ and $\zeta = 0$ becomes $\bar{V} = \mathcal{U}\bar{W}$.

The matrix of normalisation factors of the different fields that are connected to the LMs directed toward the interface $\bar{W}^{\text{toward}}$ is constructed from the amplitudes of the LMs on the left or the right side of the interface that have negative or positive group-velocity, respectively, as

$$\bar{W}^{\text{toward}} = \bar{W}_{L/R}\,\sigma^{in}_{L/R}, \tag{4.4}$$

with, for example on the left side, $\bar{\vec{W}}^{\text{toward}\,\alpha} = \bar{W}_L\,\vec{\sigma}^{in}_L{}^\alpha$, the linear combination of the amplitudes of LMs that have their group-velocity directed toward the interface. Similarly,

$$\bar{W}^{\text{away}} = \bar{W}_{L/R}\,\sigma^{out}_{L/R}, \tag{4.5}$$

wherefrom

$$\bar{W}^{\text{toward}} = \sigma^{in}_{L/R}{}^T\,\sigma^{out}_{L/R}{}^{T-1}\,\bar{W}^{\text{away}\,T}, \tag{4.6}$$

where we have used the relation between the V and W matrices, and called on the fact that, them being bases sets, the uniqueness of solutions implies that if they transform at a specific point ($\zeta = 0$ in 4.5), they must do so at any point. In Eq. (4.6), we have related the amplitude of the incoming field to that of the outgoing field by means of the scattering matrix S, with

$$S^T = \sigma^{in}_L{}^T\,\sigma^{out}_L{}^{T-1} = \sigma^{in}_R{}^T\,\sigma^{out}_R{}^{T-1}. \tag{4.7}$$

It appears that, in order to calculate the scattering matrix, all that needs being done is to calculate the above σ matrices. These are 8×8 matrices whose components are the coefficient of each LM in the linear expansion of each GM. Thus, they are calculated by using the matching conditions.

4.1.2   Matching Local Amplitudes to Calculate Global Ones

In terms of the formalism introduced in the previous paragraph, the fields on the left and on the right of the interface are related by the matching conditions

$$\bar{W}_L\,\sigma^{in}_L{}^\alpha = \bar{W}_R\,\sigma^{in}_R{}^\alpha, \tag{4.8}$$

for an *in* field α, and

$$\bar{W}_L \, \sigma_L^{out \, \alpha} = \bar{W}_R \, \sigma_R^{out \, \alpha}, \tag{4.9}$$

for an *out* field α. In Eqs. (4.8) and (4.9), the σ^α are 8×8 matrices. For every one of these matrices, there are a further 7 constraints to the 8 matching conditions (4.1)[1]:

- when defining an *in* GM, we set the amplitude of the other LMs that propagate toward the interface 0;
- under wavepacket normalisation, the defining input LM can be regarded as having a finite and tiny bandwidth—i.e., for negative times this LM is the only existing LM and has to be normalised with respect to itself. Thus the defining LM has unit amplitude.

We now proceed to calculating the σ matrices in each mode configuration. Then, each column in Eq. (4.8) can be written in terms of 8 dimensional column vectors $\vec{\sigma}^\alpha$—

$$\vec{\sigma}_L^{in \, \alpha} = A \vec{\sigma}_R^{in \, \alpha}. \tag{4.10}$$

We define the matrix A—that is composed of the product amplitudes of LMs on either side of the interface—as

$$A = \bar{W}_L^{-1} \, \bar{W}_R. \tag{4.11}$$

It is possible to calculate the σ matrices in terms of the elements of the A matrix for each mode configuration. We will now study two such mode configurations in detail, which will culminate in explicitly deriving the *in* and *out* σ matrices, yielding the S matrix.

4.1.2.1 Example 1: Mode Configuration 3—Disturbance in the Gravitational Field

We arrange both the global and local modes in decreasing order of comoving frame wavenumber k': *u uo mo lo l nl no nu*. We use matrices to relate GMs (columns) to LMs (rows), whereby the first column (row) of a matrix describes the GM (LM) *u*, the second *uo* and so on. In mode configuration 3, there are 8 oscillating LMs on either side of the interface. Then, (4.10) reads

---

[1]For the unphysical (exponentially growing) mode, this is different: it is defined as the unphysical mode *only* on one side. This GM serves as *in*—and identically as *out*—mode. Hence unphysical GMs scatter into themselves, by definition.

$$
\begin{pmatrix} & & & & & & & \\ & & 0\,0\,1\,0\,0\,0\,0\,0 & & & & \\ & & & & & & & \end{pmatrix} = A \begin{pmatrix} 1\,0\,0\,0\,0\,0\,0\,0 \\ 0\,1\,0\,0\,0\,0\,0\,0 \\ 0\,0\,0\,1\,0\,0\,0\,0 \\ 0\,0\,0\,0\,1\,0\,0\,0 \\ 0\,0\,0\,0\,0\,1\,0\,0 \\ 0\,0\,0\,0\,0\,0\,1\,0 \\ 0\,0\,0\,0\,0\,0\,0\,1 \end{pmatrix} \tag{4.12}
$$

$$
\Rightarrow \vec{e}_1\vec{\sigma}_{L_1}^{\,T} + \vec{e}_2\vec{\sigma}_{L_2}^{\,T} + \vec{e}_3\vec{e}_3^{\,T} + \vec{e}_4\vec{\sigma}_{L_4}^{\,T} + \ldots + \vec{e}_8\vec{\sigma}_{L_8}^{\,T} =
$$
$$
A\left(\vec{e}_1\vec{e}_1^{\,T} + \vec{e}_2\vec{e}_2^{\,T} + \vec{e}_3\vec{\sigma}_{R_3}^{\,T} + \vec{e}_4\vec{e}_4^{\,T} + \ldots + \vec{e}_8\vec{e}_8^{\,T}\right).
$$

There are 64 unknowns, materialised as "empty" components of the matrices. In (4.12) we have rewritten the matrix product of the first line in terms of the addition of the product of the vectors

$$
\vec{e}_1^{\,T} = (1\,0\,0\,0\,0\,0\,0\,0),\ \vec{e}_2^{\,T} = (0\,1\,0\,0\,0\,0\,0\,0),\ \ldots \vec{e}_8^{\,T} = (0\,0\,0\,0\,0\,0\,0\,1) \tag{4.13}
$$

with the *ith* row-vectors $\sigma_{L_{ij}}^{in} = \vec{\sigma}_{L_i}$. In order to find the σ^{in} matrix in this mode configuration, we proceed to re-arranging (4.12):

$$
\begin{pmatrix} \vec{\sigma}_{L1}^{\,T} \\ \vec{\sigma}_{L2}^{\,T} \\ 0 \\ \vec{\sigma}_{L4}^{\,T} \\ \vec{\sigma}_{L5}^{\,T} \\ \vec{\sigma}_{L6}^{\,T} \\ \vec{\sigma}_{L7}^{\,T} \\ \vec{\sigma}_{L8}^{\,T} \end{pmatrix} - \begin{pmatrix} A_{13}\,\vec{\sigma}_{R3}^{\,T} \\ A_{23}\,\vec{\sigma}_{R3}^{\,T} \\ A_{33}\,\vec{\sigma}_{R3}^{\,T} \\ A_{43}\,\vec{\sigma}_{R3}^{\,T} \\ A_{53}\,\vec{\sigma}_{R3}^{\,T} \\ A_{63}\,\vec{\sigma}_{R3}^{\,T} \\ A_{73}\,\vec{\sigma}_{R3}^{\,T} \\ A_{83}\,\vec{\sigma}_{R3}^{\,T} \end{pmatrix} = \begin{pmatrix} A_{11} & A_{12} & 0 & A_{14} & A_{15} & A_{16} & A_{17} & A_{18} \\ A_{21} & A_{22} & 0 & A_{24} & A_{25} & A_{26} & A_{27} & A_{28} \\ A_{31} & A_{32} & -1 & A_{34} & A_{35} & A_{36} & A_{37} & A_{38} \\ A_{41} & A_{42} & 0 & A_{44} & A_{45} & A_{46} & A_{47} & A_{48} \\ A_{51} & A_{52} & 0 & A_{54} & A_{55} & A_{56} & A_{57} & A_{58} \\ A_{61} & A_{62} & 0 & A_{64} & A_{65} & A_{66} & A_{67} & A_{68} \\ A_{71} & A_{72} & 0 & A_{74} & A_{75} & A_{76} & A_{77} & A_{78} \\ A_{81} & A_{82} & 0 & A_{84} & A_{85} & A_{86} & A_{87} & A_{88} \end{pmatrix} =
$$

$$
\left[\begin{pmatrix} 1\,0\,0\,0\,0\,0\,0\,0 \\ 0\,1\,0\,0\,0\,0\,0\,0 \\ 0\,0\,0\,0\,0\,0\,0\,0 \\ 0\,0\,0\,1\,0\,0\,0\,0 \\ 0\,0\,0\,0\,1\,0\,0\,0 \\ 0\,0\,0\,0\,0\,1\,0\,0 \\ 0\,0\,0\,0\,0\,0\,1\,0 \\ 0\,0\,0\,0\,0\,0\,0\,1 \end{pmatrix} - (\vec{0}\,\vec{0}\,\vec{A}_3\,\vec{0}\,\vec{0}\,\vec{0}\,\vec{0}\,\vec{0}) \right] \times \begin{pmatrix} \vec{\sigma}_{L1}^{\,T} \\ \vec{\sigma}_{L2}^{\,T} \\ \vec{\sigma}_{R3}^{\,T} \\ \vec{\sigma}_{L4}^{\,T} \\ \vec{\sigma}_{L5}^{\,T} \\ \vec{\sigma}_{L6}^{\,T} \\ \vec{\sigma}_{L7}^{\,T} \\ \vec{\sigma}_{L8}^{\,T} \end{pmatrix}.
$$

$$
\tag{4.14}
$$

In (4.14), the second line can be written as $\left(\vec{e}_1\ \vec{e}_2\ \vec{A}_3\ \vec{e}_4\ \vec{e}_5\ \vec{e}_6\ \vec{e}_7\ \vec{e}_8 \right) \times \sigma^{in}$, with σ^{in} the matrix we presently seek. To obtain it, we multiply from the left the 8×8 matrix on the right-hand-side of the first line of (4.14) with

$$
(\vec{e}_1\ \vec{e}_2\ \vec{e}_3\ \vec{e}_4\ \vec{e}_5\ \vec{e}_6\ \vec{e}_7\ \vec{e}_8)^{-1} =
\begin{pmatrix}
1 & 0 & -\frac{A_{13}}{A_{33}} & 0 & 0 & 0 & 0 & 0 \\
0 & 1 & -\frac{A_{23}}{A_{33}} & 0 & 0 & 0 & 0 & 0 \\
0 & 0 & -\frac{1}{A_{33}} & 0 & 0 & 0 & 0 & 0 \\
0 & 0 & -\frac{A_{43}}{A_{33}} & 1 & 0 & 0 & 0 & 0 \\
0 & 0 & -\frac{A_{53}}{A_{33}} & 0 & 1 & 0 & 0 & 0 \\
0 & 0 & -\frac{A_{63}}{A_{33}} & 0 & 0 & 1 & 0 & 0 \\
0 & 0 & -\frac{A_{73}}{A_{33}} & 0 & 0 & 0 & 1 & 0 \\
0 & 0 & -\frac{A_{83}}{A_{33}} & 0 & 0 & 0 & 0 & 1
\end{pmatrix},
\tag{4.15}
$$

and obtain

$$
\sigma^{in} =
\begin{pmatrix}
A_{11} - \frac{A_{13}A_{31}}{A_{33}} & A_{12} - \frac{A_{13}A_{32}}{A_{33}} & \frac{A_{13}}{A_{33}} & A_{14} - \frac{A_{13}A_{34}}{A_{33}} & - & - & - & - \\
A_{21} - \frac{A_{23}A_{31}}{A_{33}} & A_{22} - \frac{A_{23}A_{32}}{A_{33}} & \frac{A_{23}}{A_{33}} & A_{24} - \frac{A_{23}A_{34}}{A_{33}} & - & - & - & - \\
-\frac{A_{31}}{A_{33}} & -\frac{A_{32}}{A_{33}} & \frac{1}{A_{33}} & -\frac{A_{34}}{A_{33}} & - & - & - & - \\
A_{41} - \frac{A_{43}A_{31}}{A_{33}} & A_{42} - \frac{A_{43}A_{32}}{A_{33}} & \frac{A_{43}}{A_{33}} & A_{44} - \frac{A_{43}A_{34}}{A_{33}} & - & - & - & - \\
| & | & | & | & | & | & | & | \\
A_{81} - \frac{A_{83}A_{31}}{A_{33}} & A_{82} - \frac{A_{83}A_{32}}{A_{33}} & \frac{A_{83}}{A_{33}} & A_{84} - \frac{A_{83}A_{34}}{A_{33}} & - & - & - & -
\end{pmatrix}.
\tag{4.16}
$$

For the *out* modes, (4.12) is

$$
\begin{pmatrix}
1 & 0 & 0 & 0 & 0 & 0 & 0 & 0 \\
0 & 1 & 0 & 0 & 0 & 0 & 0 & 0 \\
\\
0 & 0 & 0 & 1 & 0 & 0 & 0 & 0 \\
0 & 0 & 0 & 0 & 1 & 0 & 0 & 0 \\
0 & 0 & 0 & 0 & 0 & 1 & 0 & 0 \\
0 & 0 & 0 & 0 & 0 & 0 & 1 & 0 \\
0 & 0 & 0 & 0 & 0 & 0 & 0 & 1
\end{pmatrix} = A
\begin{pmatrix}
& & 0\ 0\ 1\ 0\ 0\ 0\ 0\ 0 & \\
& & &
\end{pmatrix},
\tag{4.17}
$$

and similar algebra to the above (exchange A and A^{-1}) leads to the conclusion that

$$
\sigma_R^{out\,T} = \sigma_L^{in\,T},
\tag{4.18}
$$

that is, one can be calculated from the other by using either A or its inverse. From the above $\sigma^{in/out}$ matrices, we identify the $\sigma_{L/R}^{in/out}$ matrices:

$$\sigma_L^{in} = \begin{pmatrix} A_{11} - \frac{A_{13}A_{31}}{A_{33}} & A_{12} - \frac{A_{13}A_{32}}{A_{33}} & \frac{A_{13}}{A_{33}} & A_{14} - \frac{A_{13}A_{34}}{A_{33}} & - & - & - & - \\ A_{21} - \frac{A_{23}A_{31}}{A_{33}} & A_{22} - \frac{A_{23}A_{32}}{A_{33}} & \frac{A_{23}}{A_{33}} & A_{24} - \frac{A_{23}A_{34}}{A_{33}} & - & - & - & - \\ 0 & 0 & 1 & 0 & 0 & 0 & 0 & 0 \\ A_{41} - \frac{A_{43}A_{31}}{A_{33}} & A_{42} - \frac{A_{43}A_{32}}{A_{33}} & \frac{A_{43}}{A_{33}} & A_{44} - \frac{A_{43}A_{34}}{A_{33}} & - & - & - & - \\ | & | & | & | & | & | & | & | \\ A_{81} - \frac{A_{83}A_{31}}{A_{33}} & A_{82} - \frac{A_{83}A_{32}}{A_{33}} & \frac{A_{83}}{A_{33}} & A_{84} - \frac{A_{83}A_{34}}{A_{33}} & - & - & - & - \end{pmatrix}$$

$$\sigma_R^{in} = \begin{pmatrix} 1 & 0 & 0 & 0 & 0 & 0 & 0 & 0 \\ 0 & 1 & 0 & 0 & 0 & 0 & 0 & 0 \\ -\frac{A_{31}}{A_{33}} & -\frac{A_{32}}{A_{33}} & \frac{1}{A_{33}} & -\frac{A_{34}}{A_{33}} & -\frac{A_{35}}{A_{33}} & -\frac{A_{36}}{A_{33}} & -\frac{A_{37}}{A_{33}} & -\frac{A_{38}}{A_{33}} \\ 0 & 0 & 0 & 1 & 0 & 0 & 0 & 0 \\ 0 & 0 & 0 & 0 & 1 & 0 & 0 & 0 \\ 0 & 0 & 0 & 0 & 0 & 1 & 0 & 0 \\ 0 & 0 & 0 & 0 & 0 & 0 & 1 & 0 \\ 0 & 0 & 0 & 0 & 0 & 0 & 0 & 1 \end{pmatrix}$$

$$\sigma_L^{out} = \begin{pmatrix} 1 & 0 & 0 & 0 & 0 & 0 & 0 & 0 \\ 0 & 1 & 0 & 0 & 0 & 0 & 0 & 0 \\ -\frac{A_{31}^{-1}}{A_{33}^{-1}} & -\frac{A_{32}^{-1}}{A_{33}^{-1}} & \frac{1}{A_{33}^{-1}} & -\frac{A_{34}^{-1}}{A_{33}^{-1}} & -\frac{A_{35}^{-1}}{A_{33}^{-1}} & -\frac{A_{36}^{-1}}{A_{33}^{-1}} & -\frac{A_{37}^{-1}}{A_{33}^{-1}} & -\frac{A_{38}^{-1}}{A_{33}^{-1}} \\ 0 & 0 & 0 & 1 & 0 & 0 & 0 & 0 \\ 0 & 0 & 0 & 0 & 1 & 0 & 0 & 0 \\ 0 & 0 & 0 & 0 & 0 & 1 & 0 & 0 \\ 0 & 0 & 0 & 0 & 0 & 0 & 1 & 0 \\ 0 & 0 & 0 & 0 & 0 & 0 & 0 & 1 \end{pmatrix} \quad (4.19)$$

$$\sigma_R^{out} = \begin{pmatrix} A_{11}^{-1} - \frac{A_{13}^{-1}A_{31}^{-1}}{A_{33}^{-1}} & A_{12}^{-1} - \frac{A_{13}^{-1}A_{32}^{-1}}{A_{33}^{-1}} & \frac{A_{13}^{-1}}{A_{33}^{-1}} & A_{14}^{-1} - \frac{A_{13}^{-1}A_{34}^{-1}}{A_{33}^{-1}} & - & - & - & - \\ A_{21}^{-1} - \frac{A_{23}^{-1}A_{31}^{-1}}{A_{33}^{-1}} & A_{22}^{-1} - \frac{A_{23}^{-1}A_{32}^{-1}}{A_{33}^{-1}} & \frac{A_{23}^{-1}}{A_{33}^{-1}} & A_{24}^{-1} - \frac{A_{23}^{-1}A_{34}^{-1}}{A_{33}^{-1}} & - & - & - & - \\ 0 & 0 & 1 & 0 & 0 & 0 & 0 & 0 \\ A_{41}^{-1} - \frac{A_{43}^{-1}A_{31}^{-1}}{A_{33}^{-1}} & A_{42}^{-1} - \frac{A_{43}^{-1}A_{32}^{-1}}{A_{33}^{-1}} & \frac{A_{43}^{-1}}{A_{33}^{-1}} & A_{44}^{-1} - \frac{A_{43}^{-1}A_{34}^{-1}}{A_{33}^{-1}} & - & - & - & - \\ | & | & | & | & | & | & | & | \\ A_{81}^{-1} - \frac{A_{83}^{-1}A_{31}^{-1}}{A_{33}^{-1}} & A_{82}^{-1} - \frac{A_{83}^{-1}A_{32}^{-1}}{A_{33}^{-1}} & \frac{A_{83}^{-1}}{A_{33}^{-1}} & A_{84}^{-1} - \frac{A_{83}^{-1}A_{34}^{-1}}{A_{33}^{-1}} & - & - & - & - \end{pmatrix}.$$

Thus

$$
\sigma_L^{out\,T} =
\begin{pmatrix}
1 & 0 & -\frac{A_{31}^{-1}}{A_{33}^{-1}} & 0 & 0 & 0 & 0 & 0 \\
0 & 1 & -\frac{A_{32}^{-1}}{A_{33}^{-1}} & 0 & 0 & 0 & 0 & 0 \\
0 & 0 & -\frac{1}{A_{33}^{-1}} & 0 & 0 & 0 & 0 & 0 \\
0 & 0 & -\frac{A_{34}^{-1}}{A_{33}^{-1}} & 1 & 0 & 0 & 0 & 0 \\
0 & 0 & -\frac{A_{35}^{-1}}{A_{33}^{-1}} & 0 & 1 & 0 & 0 & 0 \\
0 & 0 & -\frac{A_{36}^{-1}}{A_{33}^{-1}} & 0 & 0 & 1 & 0 & 0 \\
0 & 0 & -\frac{A_{37}^{-1}}{A_{33}^{-1}} & 0 & 0 & 0 & 1 & 0 \\
0 & 0 & -\frac{A_{38}^{-1}}{A_{33}^{-1}} & 0 & 0 & 0 & 0 & 1
\end{pmatrix}
\Rightarrow
\sigma_L^{out\,T^{-1}} =
\begin{pmatrix}
1 & 0 & A_{31}^{-1} & 0 & 0 & 0 & 0 & 0 \\
0 & 1 & A_{32}^{-1} & 0 & 0 & 0 & 0 & 0 \\
1 & 0 & A_{33}^{-1} & 0 & 0 & 0 & 0 & 0 \\
0 & 0 & A_{34}^{-1} & 1 & 0 & 0 & 0 & 0 \\
0 & 0 & A_{35}^{-1} & 0 & 1 & 0 & 0 & 0 \\
0 & 0 & A_{36}^{-1} & 0 & 0 & 1 & 0 & 0 \\
0 & 0 & A_{37}^{-1} & 0 & 0 & 0 & 1 & 0 \\
0 & 0 & A_{38}^{-1} & 0 & 0 & 0 & 0 & 1
\end{pmatrix}
\tag{4.20}
$$

and

$$
\sigma_L^{in\,T} =
\begin{pmatrix}
A_{11} - \frac{A_{13}A_{31}}{A_{33}} & A_{12} - \frac{A_{13}A_{32}}{A_{33}} & 0 & A_{14} - \frac{A_{13}A_{34}}{A_{33}} & - & - & - & - \\
A_{21} - \frac{A_{23}A_{31}}{A_{33}} & A_{22} - \frac{A_{23}A_{32}}{A_{33}} & 0 & A_{24} - \frac{A_{23}A_{34}}{A_{33}} & - & - & - & - \\
\frac{A_{13}}{A_{33}} & \frac{A_{23}}{A_{33}} & 1 & \frac{A_{43}}{A_{33}} & \frac{A_{53}}{A_{33}} & \frac{A_{63}}{A_{33}} & \frac{A_{73}}{A_{33}} & \frac{A_{83}}{A_{33}} \\
A_{41} - \frac{A_{43}A_{31}}{A_{33}} & A_{42} - \frac{A_{43}A_{32}}{A_{33}} & 0 & A_{44} - \frac{A_{43}A_{34}}{A_{33}} & - & - & - & - \\
| & | & | & | & | & | & | & | \\
A_{81} - \frac{A_{83}A_{31}}{A_{33}} & A_{82} - \frac{A_{83}A_{32}}{A_{33}} & 0 & A_{84} - \frac{A_{83}A_{34}}{A_{33}} & - & - & - & -
\end{pmatrix}.
\tag{4.21}
$$

Finally, by (4.7), we obtain the scattering matrix when there are 8 oscillatory mode-solutions on either side of the interface,

$$
S_{8\times8} =
\begin{pmatrix}
A_{11} - \frac{A_{13}A_{31}}{A_{33}} & A_{21} - \frac{A_{23}A_{31}}{A_{33}} & -\frac{A_{31}}{A_{33}} & A_{41} - \frac{A_{43}A_{31}}{A_{33}} & - & - & - & - \\
A_{12} - \frac{A_{13}A_{32}}{A_{33}} & A_{22} - \frac{A_{23}A_{32}}{A_{33}} & -\frac{A_{32}}{A_{33}} & A_{42} - \frac{A_{43}A_{32}}{A_{33}} & - & - & - & - \\
\frac{A_{13}}{A_{33}} & \frac{A_{23}}{A_{33}} & \frac{1}{A_{33}} & \frac{A_{43}}{A_{33}} & - & - & - & - \\
A_{14} - \frac{A_{13}A_{34}}{A_{33}} & A_{24} - \frac{A_{23}A_{34}}{A_{33}} & -\frac{A_{34}}{A_{33}} & A_{44} - \frac{A_{43}A_{34}}{A_{33}} & - & - & - & - \\
| & | & | & | & | & | & | & | \\
A_{18} - \frac{A_{13}A_{38}}{A_{33}} & A_{28} - \frac{A_{23}A_{38}}{A_{33}} & -\frac{A_{38}}{A_{33}} & A_{48} - \frac{A_{43}A_{38}}{A_{33}} & - & - & - & -
\end{pmatrix}
\tag{4.22}
$$

In (4.22), we have completed the derivation of the S matrix for mode configuration 3, the frequency interval over which there are 8 oscillatory solutions to the field equations on either side of the interface, in terms of the amplitudes of the LMs on either side of the interface. This derivation followed from the matching conditions for the fields and their first spatial derivative at the interface and results in a straight-forward expression that can easily be implemented—in Mathematica for the sake of this Thesis and [15].

4.1.2.2 Example 2: Mode Configuration 4—the Black Hole

We will now perform the same steps as those detailed in the above paragraph for the 4th mode configuration found in Sect. 3.2.3. Over the frequency interval of interest, the RIF acts as a black hole horizon to modes of the field: light cannot propagate from the region on the left of the interface to the region on the right as motion is only possible in one direction in the left-hand-side region, whilst motion is possible in both ζ directions in the right-hand-side region. We found that there are 8 oscillatory solutions in the RHS region (with a single mode, *mo*, allowing light to propagate to the right, away from the interface), and 6 oscillatory and 2 complex solutions in the LHS region (with all oscillary solutions having negative comoving group velocity). In this situation, the *in*-modes define A as

$$\begin{pmatrix} 0 \\ 0 \\ 0\,0\,1\,0\,0\,0\,0\,0 \\ 0 \\ 0 \\ 0 \\ 0 \\ 0 \end{pmatrix} = A \begin{pmatrix} 1\,0\ 0\,0\,0\,0\,0 \\ 0\,1\ 0\,0\,0\,0\,0 \\ \\ 0\,0\ 1\,0\,0\,0\,0 \\ 0\,0\ 0\,1\,0\,0\,0 \\ 0\,0\ 0\,0\,1\,0\,0 \\ 0\,0\ 0\,0\,0\,1\,0 \\ 0\,0\ 0\,0\,0\,0\,1 \end{pmatrix} \tag{4.23}$$

in which the global (columns) and local (rows) modes are sorted in decreasing order moving frame wavenumber k': $u\ uo\ g_L\ lo\ l\ nl\ no\ nu$ for the GMs and LMs on the left of the interface (first matrix in 4.23), and $u\ uo\ g_L\ lo\ l\ nl\ no\ nu$ for the GMs and $u\ uo\ mo\ lo\ l\ nl\ no\ nu$ for the LMs on the right of the interface (third matrix in 4.23). Conversely, the *out*-modes define A as

$$\begin{pmatrix} 1\,0\ 0\,0\,0\,0\,0 \\ 0 \\ 0\,1\ 0\,0\,0\,0\,0 \\ 0\,0\ 1\,0\,0\,0\,0 \\ 0\,0\ 0\,1\,0\,0\,0 \\ 0\,0\ 0\,0\,1\,0\,0 \\ 0\,0\ 0\,0\,0\,1\,0 \\ 0\,0\ 0\,0\,0\,0\,1 \end{pmatrix} = A \begin{pmatrix} \\ \\ 0\ 1\,0\,0\,0\,0\,0 \\ \\ \\ \\ \\ \end{pmatrix} \tag{4.24}$$

with the GMs and LMs ordered as in (4.24). Similar algebra to that used in the first example 4.1.2.1 allows to find the σ-matrices for *out*-modes on the left of the interface:

$$\sigma_L^{out} = \begin{pmatrix} 1 & 0 & 0 & 0 & 0 & 0 & 0 & 0 \\ -\frac{A_{31}^{-1}}{A_{33}^{-1}} & 0 & \frac{1}{A_{33}^{-1}} & -\frac{A_{34}^{-1}}{A_{33}^{-1}} & -\frac{A_{35}^{-1}}{A_{33}^{-1}} & -\frac{A_{36}^{-1}}{A_{33}^{-1}} & -\frac{A_{37}^{-1}}{A_{33}^{-1}} & -\frac{A_{38}^{-1}}{A_{33}^{-1}} \\ 0 & 1 & 0 & 0 & 0 & 0 & 0 & 0 \\ 0 & 0 & 0 & 1 & 0 & 0 & 0 & 0 \\ 0 & 0 & 0 & 0 & 1 & 0 & 0 & 0 \\ 0 & 0 & 0 & 0 & 0 & 1 & 0 & 0 \\ 0 & 0 & 0 & 0 & 0 & 0 & 1 & 0 \\ 0 & 0 & 0 & 0 & 0 & 0 & 0 & 1 \end{pmatrix}$$

$$\Rightarrow \sigma_L^{out\,T\,-1} = \begin{pmatrix} 1 & 0 & A_{31}^{-1} & 0 & 0 & 0 & 0 & 0 \\ 0 & 0 & A_{32}^{-1} & 0 & 0 & 0 & 0 & 0 \\ 0 & 1 & 0 & 0 & 0 & 0 & 0 & 0 \\ 0 & 0 & A_{34}^{-1} & 1 & 0 & 0 & 0 & 0 \\ 0 & 0 & A_{35}^{-1} & 0 & 1 & 0 & 0 & 0 \\ 0 & 0 & A_{36}^{-1} & 0 & 0 & 1 & 0 & 0 \\ 0 & 0 & A_{37}^{-1} & 0 & 0 & 0 & 1 & 0 \\ 0 & 0 & A_{38}^{-1} & 0 & 0 & 0 & 0 & 1 \end{pmatrix}. \tag{4.25}$$

For the *in*-modes on the left of the interface we find

$$\sigma_L^{in\,T} = \begin{pmatrix} A_{11} - \frac{A_{13}A_{31}}{A_{33}} & A_{12} - \frac{A_{13}A_{32}}{A_{33}} & 0 & A_{14} - \frac{A_{13}A_{34}}{A_{33}} & - & - & - & - \\ A_{21} - \frac{A_{23}A_{31}}{A_{33}} & A_{22} - \frac{A_{23}A_{32}}{A_{33}} & 0 & A_{24} - \frac{A_{23}A_{34}}{A_{33}} & - & - & - & - \\ 0 & 0 & 1 & 0 & 0 & 0 & 0 & 0 \\ A_{41} - \frac{A_{43}A_{34}}{A_{33}} & A_{42} - \frac{A_{43}A_{32}}{A_{33}} & 0 & A_{44} - \frac{A_{43}A_{34}}{A_{33}} & - & - & - & - \\ | & | & | & | & | & | & | & | \\ A_{81} - \frac{A_{83}A_{31}}{A_{33}} & A_{82} - \frac{A_{83}A_{32}}{A_{33}} & 0 & A_{84} - \frac{A_{83}A_{34}}{A_{33}} & - & - & - & - \end{pmatrix}. \tag{4.26}$$

Wherefrom the scattering matrix in mode configuration 2—in which there are 6 and 8 oscillatory mode-solutions on the left and on the right of the interface, respectively—is

$$S_{6\times 8} = \begin{pmatrix} A_{11} - \frac{A_{13}A_{31}}{A_{33}} & 0 & -\frac{A_{31}}{A_{33}} & A_{41} - \frac{A_{43}A_{31}}{A_{33}} & - & - & - & - \\ A_{12} - \frac{A_{13}A_{32}}{A_{33}} & 0 & -\frac{A_{32}}{A_{33}} & A_{42} - \frac{A_{43}A_{32}}{A_{33}} & - & - & - & - \\ 0 & 1 & 0 & 0 & 0 & 0 & 0 & 0 \\ A_{14} - \frac{A_{13}A_{34}}{A_{33}} & 0 & -\frac{A_{34}}{A_{33}} & A_{44} - \frac{A_{43}A_{34}}{A_{33}} & - & - & - & - \\ | & | & | & | & | & | & | & | \\ A_{18} - \frac{A_{13}A_{38}}{A_{33}} & 0 & -\frac{A_{38}}{A_{33}} & A_{48} - \frac{A_{43}A_{38}}{A_{33}} & - & - & - & - \end{pmatrix}. \tag{4.27}$$

As in the previous example, we have worked from the matching conditions for the fields and their first spatial derivatives to match the amplitudes of the GMs (built from the LMs—solutions to the dispersion relation on either side of the interface). We thus algebraically derived the scattering matrix that describes mode mixing over the interval in which the interface acts as a black hole horizon. Glancing back on Eqs. (4.22) and (4.27), we remark that S has a block matrix form, with four partitions arranged around the 3rd row and 3rd column for (4.22) and 2nd column and 3rd row for (4.27). This form, which is an intrinsic property of the construction (ordering of the GMs and LMs) of the S-matrix in our algebra, accounts for the non-coupling of the oscillatory GMs to the non-oscillatory GMs. It is thus a property that one would test for when checking the numerical calculation of the S-matrix upon calculating spectra—as we will do in the next section of this dissertation.

4.1.2.3 Quasi-unitarity of the Scattering Matrix

In performing the algebra toward the scattering matrix in mode configurations 3 and 4, we have encountered the main and usual steps of our algorithm: first we write the A-matrix that gathers the amplitude of the *in*- and *out*-modes on either side of the interface, second we re-arrange the matrix equations to clearly identify the components of the σ-matrices, we then read off the elements of these matrices on the left and the right of the interface and finally use the expression of the scattering matrix as a function of these σ-matrices to explicitly derive it. We found that the scattering matrix is, by construction, a block matrix arranged around a row and a column that account for the non-coupling of oscillatory GMs with non-oscillatory GMs.

The "normalised" scattering matrix implemented in the algorithm presented in this section transforms *in* GMs into *out* GMs by Eq. (3.80). The GMs are normalised by Eq. (3.84) and, as a result of this normalisation, the scattering matrix is a quasi-unitary matrix. This can be seen by studying the conservation of the probability current density j (see Eq. 3.62) across the interface: the matching conditions across the interface imply that

$$-iu(A_L^* \Pi_{A_L^*} + \sum_{i=1}^{3} P_{i_L}^* \Pi_{P_{i_L}^*} - \Pi_{A_L} A_L - \sum_{i=1}^{3} \Pi_{P_{i_L}} P_{i_L}) =$$

$$-iu(A_R^* \Pi_{A_R^*} + \sum_{i=1}^{3} P_{i_R}^* \Pi_{P_{i_R}^*} - \Pi_{A_R} A_R - \sum_{i=1}^{3} \Pi_{P_{i_R}} P_{i_R}) \qquad (4.28)$$

$$\rightarrow j_L = j_R,$$

where the current and fields on the left (right) of the interface have been ascribed a subfix L (R). Rearranging (4.28) yields

$$
\begin{aligned}
V^{out\dagger} g\ V^{out} &= V^{in\dagger} g\ V^{in} \\
V^{in\dagger} S^{\dagger} g\ S\ V^{in} &= V^{in\dagger} g\ V^{in} \\
\to\ S^{\dagger} g\ S &= g,
\end{aligned}
\tag{4.29}
$$

where g is the diagonal matrix with N_+ diagonal elements equal to 1 and N_- equal to -1, with N_+ and N_- the number of GMs of positive and negative norm, respectively. In the present case (8 branches dispersion relation), $g = diag(1, 1, 1, 1, 1, -1, -1, -1)$.

Since the scattering matrix is quasi-unitary, its rows obey the normalisation condition (3.84), meaning that numerically adding the amplitude squared of all components of a row (multiplied by $sign(\omega_\alpha)$, where ω_α is the laboratory frame frequency and α is the mode—$u, uo, mo...$) should yield 1. Indeed, by (4.29), S is a member of the indefinite unitary group $U(N_+, N_-)$. Thus the scattering matrix obeys $\left[S^T\right]^\star g S = g$, and hence $S^{-1} = g \left[S^T\right]^\star g$.

Now that we have learnt the steps toward deriving the scattering matrix, and understood what are the essential features of this matrix, we can simply state the scattering matrix one obtains for mode configurations 1 and 5 (6 oscillatory solutions only on either side of the interface), and 2 (8 and 6 oscillatory solutions on the LHS and RHS, respectively, of the interface—the white hole analogue).

4.2 Algorithmic of Laboratory Frame Emission

We now want to use the scattering matrix calculated in the previous section to compute the laboratory frame spectral density of emission from a dielectric step-like boundary separating two homogeneous media. For this purpose, we will create an algorithm that, for each wavelength (as measured in the laboratory frame), returns a scalar quantity: the spectral density of emission. This density might be the result of emission into a collection of any of the global modes (GMs) defined in Sect. 3.2.4 (see Eq. 3.79).

We begin our investigation with a close study of the dispersion relation in the laboratory and comoving frames, and thus identify the modes that contribute to the emission at each laboratory frame wavelength. Following on which we calculate the rate of particle production in each contributing mode at each laboratory frame wavelength, and add them (where necessary) to compute the laboratory frame spectral density (LSD). This shall allow us, in the final section of this chapter, to identify key features of spontaneous emission of light at the horizon.

4.2.1 Journeying Along the Optical Branch

Our aim is to create a function that, for a certain laboratory frame frequency, ω, calculates the contributions from all modes in which light is emitted from the boundary in the medium to the spectral density.

The spectrum that will thus be computed should be observable by some sort of apparatus at rest with respect to the medium in the laboratory. Thus, we consider emission as it can be detected from one end of the 1D medium only—say the right, in reference to the positive x axis direction. On physical grounds, this implies that light in a mode that would have negative group velocity in the laboratory frame (that would move to the left) will not be taken into account in our calculations. In other words, the spectrum will be made of contributions from modes that allow for light to propagate in the same direction as the refractive index front (RIF) in the medium. Furthermore, as in earlier calculations, we operate at frequencies for which there are no contributions from the *top* branch in the dispersion relation (3.59). In Fig. 4.2, we plot an example of such an 8 branches dispersion relation for two homogeneous media that differ in their refractive index in the laboratory frame. Following the convention used so far in this dissertation, the medium with highest refractive index (orange curves in Fig. 4.2) is the region on the left of the interface (as in Fig. 4.2). Clearly, the medium with lowest refractive index (black curves in Fig. 4.2) is then the region on the right of the interface.

We want to calculate contributions to the laboratory-frame spectral density of emission (LSD) over optical frequencies. Only oscillatory mode-solutions may contribute to the emission at a given laboratory frame frequency ω. By construction of our field theory (Sect. 3.2.4), monochromatic solutions to the field equations have positive moving-frame-frequency. These oscillatory mode solutions are found at the intersection between the $\omega' = cst$ contour line and the (positive- and negative-laboratory-frame) optical branch on either side of the interface.

We have previously established that the number of oscillatory mode-solutions is a function of the comoving frequency ω'. We found that there exists up to 5 mode

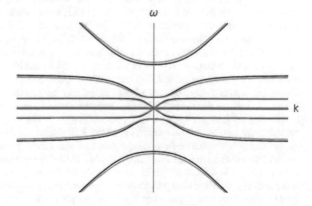

Fig. 4.2 Sellmeier dispersion relation (3.59) of the left (right) region of the RIF in orange (black) in the laboratory frame. In each medium there are three resonances (one is very close to the horizontal axis), and hence 8 branches. An increase in the refractive index distorts the branches by lowering the resonance frequencies and increasing the inertia of the excitons

configurations in the three realisable regimes of refractive index change at the interface between the homogeneous media of the RIF (as exemplified in Fig. 3.6). These mode configurations describe the variation of positive-optical-frequency oscillatory mode-solutions on either side of the interface. Indeed, there is always only one oscillatory mode-solution on all other branches of the dispersion relation. In Fig. 4.3a we show the positive-optical-frequency branch (for the two homogeneous regions around the interface) as seen in the frame co-moving with the RIF at $u = 0.66c$. The study of this diagram tells us which modes are *out* GMs at a given ω'. Modes that may contribute to the LSD are found in the interval $[\omega_{minL}, \omega_{maxR}]$.

As illustrated on Fig. 4.3c, for each ω, we find 4 intersection points of positive frequency on either side of the interface. By symmetry, we also find 4 intersection

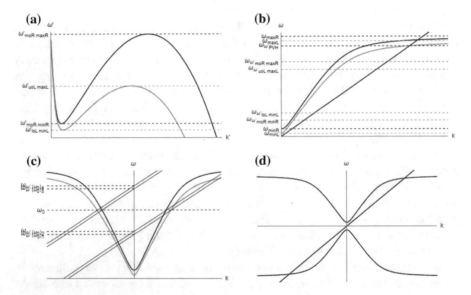

Fig. 4.3 Optical branches of the the dispersion relation in the moving and laboratory frame. **a** The turning points of the positive laboratory-frame-optical-frequency branch (as seen from the moving frame) on the left and right side of the RIF, in orange and black, respectively, define the extrema of the intervals of emission for the modes *loL, moR* and *uoL*. **b** When Doppler shifted to the laboratory frame, and because they are defined on different sides of the interface (i.e., in regions of different refractive indices), these emission intervals are not always distinct anymore: the emission interval of *moR* overlaps with those of *loL* and *uoL* at low and high ω, respectively. For frequencies higher than the Doppler shifted zero-comoving-frame-frequency ($\omega' = 0$, phase-velocity horizon condition, in blue), only *noL* contributes to the laboratory frame spectral density of emission. **c** For any laboratory-frame frequency ω_0, there are always 2 corresponding LMs, of moving-frame frequencies ($\omega'(\omega_0)_1, \omega'(\omega_0)_2, \omega'(\omega_0)_3$ and $\omega'(\omega_0)_4$, in blue) on the positive (laboratory-frame) optical-frequency branch of the RIF (black—right of the interface, orange, left of the interface)—as well as their 4 counterparts on the negative (laboratory-frame) optical-frequency branch (not shown). If these LMs define *out* GMs, the latter contribute to the emission at ω_0. Because LMs and GMs are defined for positive moving-frame frequencies only, the $\omega' = 0$ contour (in blue) separates the (positive and negative laboratory-frame frequency) optical branch in two regions, only modes belonging to the regions highlighted in purple in (**d**) contribute to the LSD

points of negative frequency on either side of the interface, for a total of 8 intersection points. The modes that contribute to the emission at ω are those that: (1) have positive comoving frequency and (2) define *out* GMs. In the left region, local modes (LMs) *noL, loL* and *uoL* define *out* GMs, whereas in the right region, only *moR* does.

Now that we have identified the modes that could contribute to optical-frequency emission, we can dwell back upon the dispersion relation plotted in the moving frame Fig. 4.3a: it is possible to find the moving-frame frequency intervals over which LMs *noL, loL, uoL* and *moR* are oscillatory solutions to the fields equations. These frequency interval limits can then be boosted back to the laboratory frame by means of the inverse Lorentz transform 3.51—we draw the Fig. 4.3b. In growing order of frequency, we find that over the interval

- $\left[\omega_{minL}, \omega_{\omega'_{moR\ minR}}\right]$, *loL* will be the sole contributor to the LSD;
- $\left[\omega_{\omega'_{moR\ minR}}, \omega_{\omega'_{loL\ minL}}\right]$, both *loL* and *moR* will contribute to the LSD;
- $\left[\omega_{\omega'_{loL\ minL}}, \omega_{\omega'_{uoL\ maxL}}\right]$, only *moR* contributes to the LSD;
- $\left[\omega_{\omega'_{uoL\ maxL}}, \omega_{\omega'_{moR\ maxR}}\right]$, *moR* and *uoL* contribute together to the LSD;
- $\left[\omega_{\omega'_{moR\ maxR}}, \omega_{PVH}\right]$, *uoL* alone contributes to the LSD;
- $\left[\omega_{\omega'_{PVH}}, \omega_{\omega'_{cut}}\right]$, laboratory frame emission will arise from contributions of the negative norm mode *noL*.

Note that ω_{PVH} is the laboratory-frame frequency for which the moving-frame frequency is $\omega' = 0$, and ω'_{cut} is the maximum moving-frame frequency for which there are no contributions from the *top* dispersion relation branch. All frequencies are shown in Fig. 4.3b. For laboratory frequencies outside of the above-stated intervals, there is no emission from light in optical modes. When $\omega' = 0$, there is a phase velocity horizon (PVH): the contour-line that would then be drawn separates two "regions" of the (positive- and negative-) laboratory-frame frequency branch—one that contribute to the emission from one that does not, see Fig. 4.3d.

To summarise, we have identified the modes that contribute to the LSD as a function of the laboratory-frame frequency. We found that emission stems from contributions of up to two modes over various intervals—and thus expect the resulting spectrum to be highly structured in those intervals. This shall later allow us to identify intervals of horizon-like emission. But first, let us progress further with the writing of our LSD function and see how to calculate the contribution from each mode.

4.2.2   Rate of Particle Production in a Mode

When computing the laboratory frame spectrum, we will input a laboratory-frame frequency ω to the function we are presently creating, and it will output the density of emission per unit time and unit bandwidth.

The algorithm created in Sects. 4.1.1–4.1.2 basically calculates the scattering matrix, i.e., mode conversion in the moving frame. We have previously explained how to find the moving-frame frequencies ω' for which a GM contributes to the laboratory spectral density (LSD) at frequency ω. Thus all that remains to be done is to implement the calculation of $I_{\omega'}^{'\alpha}$ (Eq. 3.93), with α the *out* GM that contributes to emission at ω, and to calculate the resulting rate of photon production per unit time and unit frequency in the laboratory frame. A GM's contribution to the emission at ω of emission in the laboratory frame is computed by [14]

$$I_{\omega}^{\alpha} = \left(1 - \frac{u}{v_g(\omega)} \right) I_{\omega'}^{'\alpha}, \tag{4.30}$$

where $v_g(\omega)$ is the laboratory group velocity at ω. The total spectral density at ω is then found by adding the contributions of all GMs to the emission [15]—

$$I_{\omega} = \sum_{\alpha} I_{\omega}^{\alpha}, \tag{4.31}$$

yielding the spectral density as a function of frequency. The latter converts to the spectral density as a function of wavelength by the factor $\omega^2/(2\pi c)$.

In conclusion, in this section we have worked through the algorithm that would be implemented to calculate the total spectral emission density as it can be measured in the laboratory. In the following section, we will implement this function and compute spectra of light spontaneously emitted at the RIF.

4.3 Emission Spectra and Photon Flux

In the previous two sections of this dissertation, we have devised two algorithms based on an analytical calculation: one to derive the scattering matrix that describes the scattering of an incoming field into an outgoing field, and one to calculate the emission spectra as they would be observed in the laboratory frame.

The first algorithm describes the steps to be taken to calculate the elements of the scattering matrix. It is a generic method, that is valid for any physical system that can be described by a dispersion relation that would feature up to three poles. In present Thesis we study the scattering of light at a step-like refractive index front (RIF), as schematically depicted in Fig. 3.4.

In this section of the dissertation, we will present the main numerical results of this Thesis—as they were published in 2015 in [15]: the scattering of input modes in the vacuum state (devoid of particle, photon, population). This will result in showing how light is spontaneously emitted from the vacuum at the interface (a result of quantum fluctuations of the vacuum) according to Eq. (3.93). We will study all the mode configurations found in Sect. 3.2.3 of this dissertation, and comment on the particular structure of the emission spectra over specific moving-frame-frequency intervals. Further to the findings of [15], we will present results for all modes and all refractive-index-increase magnitudes (that is, low, medium, and large δn increase under the step).

We will then proceed to computing the spectra of emission as they can be observed in the laboratory frame for realistic experimental situations.

4.3.1 Emission in the Moving Frame

We use the scattering matrix to compute the spectra of emission into all modes as seen from the moving frame. We consider light in bulk fused silica. The material resonances are $\lambda_{1,2,3} = 9904$, 116, and 68.5 nm, respectively, and the elastic constants are $\kappa_{1,2,3} = 0.07142$, 0.03246, and 0, 05540, respectively [14]. The velocity of the refractive index front (RIF) is $u = 0.66c$, corresponding to a group index of 1.5.

We first consider spectra of emission into moR, the unique right-going mode for all (small, medium and large) magnitudes of refractive index change δn.

4.3.1.1 Emission into the Uniquely Escaping Mode

Figure 4.4a displays the spectrum of emission into moR for a large, medium and small increase in the refractive index under the RIF. Spectral emission is constrained to the subluminal interval (SLI) on the right of the interface ($[\omega'_{min\ R}, \omega'_{max\ R}]$ in Fig. 3.5) where the mode moR exists. For large ($\delta n \geq 0.056$) and medium ($0.04 < \delta n < 0.056$) increase in the refractive index, an optical horizon exists over the entire right SLI. However, when δn is smaller than 0.04 (small refractive index increase), an optical horizon exists for only part of this interval (i.e., $[\omega'_{max\ L}, \omega'_{max\ R}]$ in Fig. 3.5) because at lower frequencies the SLIs of the left and right regions overlap (see Fig. 3.5). We observe that for large and medium δn, the emission spectra are quasi-thermal, with almost constant flux density over the interval of emission. In contrast, for small δn (see Fig. 4.4d, it appears that the absence of a horizon leads to a significant decrease in the emission, i.e., mode coupling, although some emission remains. We observe that the flux density drops at the extrema of the interval of emission. This decrease in the flux on the edges of the interval is due to the decrease in group-velocity of moR at these frequencies. Figure 3.5 shows of the gradient of the optical branch in the moving frame goes to zero: at $\omega' = \omega'_{max\ R}$ or $\omega' = \omega'_{min\ R}$, $\frac{\partial \omega'}{\partial k'} = 0$. Thus, moR and the interface are velocity matched, meaning that no light in moR may propagate away from the interface.

In Fig. 4.4b, we plot the emission into moR for increasing magnitudes of δn, from 0.06 to 0.12 (large refractive index change). First, we observe that the shape of the spectrum does not depend on the increase in the refractive index, only the overall magnitude of the flux increases with δn. Second, we note that for $\delta n = 0.12$ (magenta curve), we obtain exactly the same spectrum as was calculated in [14]. In Fig. 4.4c,

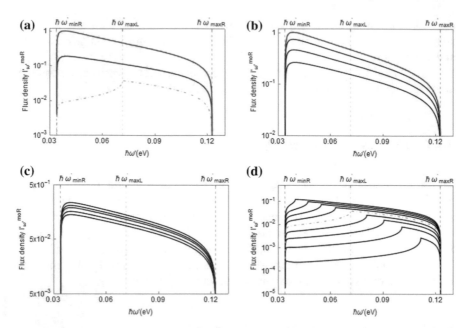

Fig. 4.4 Spectrum of emission into the uniquely right propagating mode *moR* on the right side of the RIF. The number of particles per time and bandwidth, the flux density of emission, is calculated in the moving frame of velocity $u = 0.66c$. **a** Emission is displayed for three values of δn, 2×10^{-2}, 4.9×10^{-2} and 0.12, corresponding to the regimes of low (orange dot-dashed line), medium (solid red line) and large (solid magenta line) refractive index under the step, respectively. **b** The flux density is also plotted for increasing values of δn, the large ($\delta n = 6 \times 10^{-2}$, 8×10^{-2}, 0.1, and 0.12 (solid magenta line)), **c** the medium ($\delta n = 4 \times 10^{-2}$, 4.4×10^{-2}, 4.9×10^{-2} (solid red line), 5.2×10^{-2}, 5.6×10^{-2}), and d) the low ($\delta n = 3.6 \times 10^{-2}$, 3.2×10^{-2}, 2.8×10^{-2}, 2.4×10^{-2}, 2×10^{-2} (orange dot-dashed line), 1.6×10^{-2}, 1.2×10^{-2}, 8×10^{-3}, 4×10^{-3}) regime of refractive index change

we plot the emission for medium index changes, from 0.04 to 0.056. As in the case of the large refractive change, we observe that the shape of the spectrum does not depend on δn, only the magnitude does. This is because the mode coupling does not change over the right SLI for such large increases in the refractive index.

Indeed, as was noted above, the shape of the spectrum of emission into *moR* only changes for small δn because the right and left SLIs do not overlap fully over $\left[\omega'_{min\ R}, \omega'_{max\ R} \right]$, and thus the mode coupling evolves: the nature and number of the modes that scatter into each other at the interface changes across the interval over which there is emission into *moR* and so the coupling coefficients (components of the scattering matrix (4.7)) change in nature and amplitude. In reference to Sects. 4.1.2.1 and 4.1.2.2, the relevant scattering matrix is (4.22) for $\omega' \in \left[\omega'_{min\ R}, \omega'_{max\ L} \right[$, and (4.27) for $\omega' \in \left[\omega'_{max\ L}, \omega'_{max\ R} \right]$. Remarkably, over the frequency interval in which the interface acts as a black hole event horizon to modes of the field, the shape

Fig. 4.5 Spectrum of emission as in Fig. 4.4d. To compare the shapes of the traces, spectral densities are scaled such that all traces line up with the $\delta n = 0.02$ (orange dot-dashed line). $\delta n = 5.2 \times 10^{-2}, 3.7 \times 10^{-2}, 2 \times 10^{-2}$ (orange dot-dashed line), $7 \times 10^{-3}, 1 \times 10^{-3}, 5 \times 10^{-4}$ and then range from 1.2×10^{-4} to 4×10^{-5}, in steps of to 1×10^{-5}

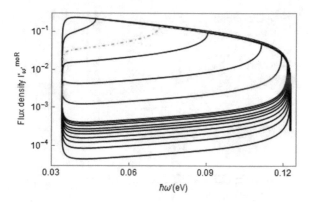

of the spectrum is independent of the refractive index change (as was observed above for $\delta n > 0.01$). This can be more clearly seen in Fig. 4.5, in which the spectra for smaller refractive index changes are scaled up to compensate the lower single-frequency rate. This is a remarkable result: all traces over orders of magnitude of index changes line up to the same shape, making it a universal signature of analogue black-hole emission. Note also that the shape differs for emission outside the black-hole-frequency interval.

It is also possible to calculate emission into modes in which light propagates on the other side of the interface (in the high refractive index region)—the following spectra are original results of this thesis and are, in their majority, presented for the first time.

4.3.1.2 Emission into All Modes for All Changes of Refractive Index

The scattering matrix also gives us the comoving flux densities of all other optical modes, of positive and negative laboratory-frame frequency (*loL, uoL* and *noL*). These modes are outgoing modes in the high refractive index region of the RIF (i.e. light in these modes propagates from the interface to the left in the moving frame). The flux density is calculated in the regimes of large, medium, and low refractive change under the step ($\delta n = 0.3$, 4.9×10^{-2}, 2×10^{-2}). As in the above study of the spectrum of *moR*, we observe that the emission is highly structured in intervals with black- or white-hole horizon, and no horizon. Remark that in the regime of large refractive index change, there is no turning point in the optical branch in the moving frame (merely an inflexion, see Fig. 3.5), thus the discrimination between modes *loL* and *uoL* in the high refractive index region becomes arbitrary. We chose to consider that mode *loL* would be the oscillatory solution at all moving-frame frequencies ω' in this case, and thus there is no emission *uoL*.

Again, the magenta lines in Fig. 4.6 reproduce the results of [14], and the discontinued blue and green, and solid purple, lines reproduce those of [15], for the large and low refractive index change regimes, respectively. The solid red lines (medium

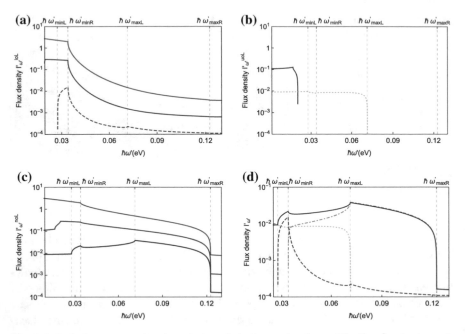

Fig. 4.6 Emission spectra of each optical mode in the moving frame. The flux density in mode *loL*, **a** *uoL*, **b** and *noL*, **c** is plotted in the regimes of large (purple line, $\delta n = 0.3$), medium (red line, $\delta n = 4.9 \times 10^{-2}$), and low ($\delta n = 2 \times 10^{-2}$) refractive index change under the step. In the regime of large refractive index change, there is no oscillatory mode *uoL* under the step. **d** For intensity comparison, all positive and negative laboratory-frame frequency modes are plotted together for $\delta n = 2 \times 10^{-2}$ (low refractive index change regime): emission into mode *noL*, purple solide line; *loL*, blue dashed line; *uoL*, green dotted line; *moR*, orange dot-dashed line

refractive index change regime) are presented here for the first time. For mode-intensity comparison purposes, we plotted the emission into all optical modes of positive and negative norm together in Fig. 4.6d for a fixed $\delta n = 0.02$. The strongest emission occurs into the optical mode with negative norm, *noL*. This emission is due to coupling with all the other positive-norm modes in the medium and is strongest where this mode that propagates in the superluminal region couples to a mode that propagates in the subluminal region (i.e., *moR*, for $\omega'_{maxL} < \omega' < \omega'_{maxR}$). In other words, the emission in *noL* is strongest over the analogue black- or white-hole intervals, when pair-wise emission with *moR* dominates—a phenomenon analogous to Hawking Radiation in black hole physics.

Mode *noL*, because it is the mode with strongest emission in the moving frame, and also because it has a negative norm, draws attention. We thus compute further emission spectra, for a variety of refractive index changes δn from the low to the large refractive index change regime (see Fig. 4.7a, as well as for the regime of medium refractive index change (see Fig. 4.7b. As for all modes, we observe that the emission spectra are highly structured in intervals of emission with black- or white-hole horizons, as well as intervals over which there is no horizons. Likewise,

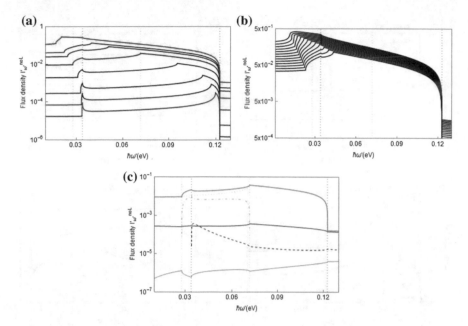

Fig. 4.7 Spectrum of emission into the optical mode of negative norm noL, in which light propagates away from the interface into the high refractive index region of the RIF. Emission is calculated in the moving frame of velocity $u = 0.66c$. **a** The flux density is displayed for increasing values of δn (4.9×10^{-2} (magenta line), $3.6 \times 10^{-2}, 3 \times 10^{-2}, 2 \times 10^{-2}$ (purple line), $1 \times 10^{-2}, 4 \times 10^{-3}, 2 \times 10^{-3}$, and 1×10^{-3}). **b** The flux is plotted for decreasing values of refractive index changes in the medium regime ($\delta n = 5.8 \times 10^{-2}, 5.6 \times 10^{-2}, 5.4 \times 10^{-2}, 5.2 \times 10^{-2}, 5 \times 10^{-2}, 4.8 \times 10^{-2}, 4.6 \times 10^{-2}, 4.4 \times 10^{-2}, 4.2 \times 10^{-2}, 4 \times 10^{-2}, 3.8 \times 10^{-2}, 3.6 \times 10^{-2}, 3.4 \times 10^{-2}$). **c** The contributions $\left|S^{noL}\tilde{\alpha}\right|^2$ of in GMs to the flux in noL are calculated for $\delta n = 2 \times 10^{-2}$. The five positive norm in GMs contribute to emission into noL: uR (black), uoR (green dotted), moL (orange dot-dashed), loR (blue dashed), and lR (brown)

the spectral width of these intervals saturates once the large refractive index regime is reached, and thus the features of the spectrum are locked in frequency and only the density of emission increases. Note that emission is increased over intervals where the interface acts as a black- or white-hole horizon.

The analytical method developed in Sect. 4.1 also allows for calculating emission into modes that lie on other branches of the dispersion relation, namely the positive- and negative- low and up laboratory-frame-frequency branches. There are 4 modes: two with positive frequency, lL and uL, and two with negative frequency, nlL and nuL. All define out GMs in which light propagates away from the interface into the high refractive index region. As noted by Finazzi and Carusotto [14], emission into those non-optical modes is significantly lower than that in optical modes. They illustrated their statement with computations of emission into the positive norm modes (lL and uL) in the large refractive index change regime—here we will present the first results for all non-optical modes of positive and negative norm in all regimes of refractive index change.

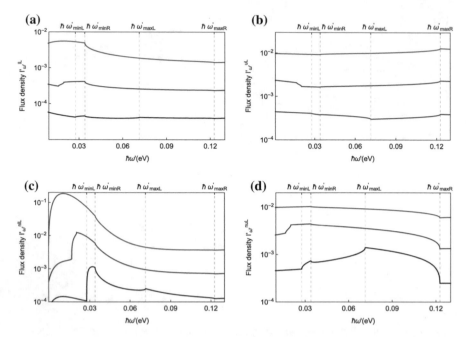

Fig. 4.8 Spectrum of emission into all non-optical modes, of positive and negative norm. Emission is calculated in the moving frame of velocity $u = 0.66c$ for a low ($\delta n = 0.02$, black line), medium ($\delta n = 0.049$, red line), and high ($\delta n = 0.3$, magenta line) change in the refractive index: emission into mode **a** lL, **b** uL, **c** nlL, and **d** nuL

In Fig. 4.8, we observe that emission in non-optical modes of all norm is at least an order of magnitude weaker than in any optical mode, for all comoving frequency ω' (see Fig. 4.6). This corroborates our earlier intuition that the study of optical modes only would reveal the essentials of horizon physics in a dispersive medium such as fused silica. Remark that, although the emission in all non-optical modes is structured into intervals with black- or white-hole horizon, and no horizon, the spectrum features vary most in the regime of low refractive index change. We note that, of all the non-optical modes, nlL (Fig. 4.8c) has the highest emission rate. This mode has a negative norm, and it would be interesting to understand why its flux density is so much higher than that in other non-optical modes (and actually relatively close to that in the weakest optical modes).

I foresee this is due to its relative "closeness" with positive-norm optical modes in the moving frame dispersion diagram (that is, it has a k', moving frame wavenumber, close to that of loL, for example). Further work should be dedicated to this question, as one might learn more about the physics of event horizon in dispersive media by shedding light upon the coupling of negative and positive norm modes across—and on the same side of—the horizon.

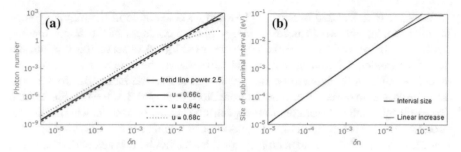

Fig. 4.9 Total emission into *moR* over the subluminal interval $\left[\omega'_{minR}, \omega'_{maxR}\right]$. **a** estimated photon number for different velocities u and **b** size of the interval in the frame moving at $u = 0.66c$ as a function of index change δn

4.3.1.3 Total Black Hole Emission

To conclude with our considerations of emission in the moving frame, it is interesting to calculate the total photon flux over the SLI $\left[\omega'_{minR}, \omega'_{maxR}\right]$ by integrating over the spectrum of Fig. 4.4d. In order to convert the flux to a realistic, although very approximate, photon number, we assume that the RIF propagates over a distance of 1mm. The resulting photon number as a function of index change δn is given in Fig. 4.9a. The number of photons excited from the vacuum first grows with power ≈ 2.5 of δn until $\delta n = 0.04$. The emission spectrum becomes wider in a linear way, as shown in Fig. 4.9b. Thus the emission rate for a single mode increases with $\delta n^{3/2}$. This scaling factor is unexplained and would deserve to be investigated further. As we explained earlier, in the regime of medium and large refractive index change, i.e., for $\delta n \geq 0.04$, the spectral width saturates, and the emission rate grows slower accordingly. However, these index steps are difficult to reach experimentally by nonlinear pulses. The rate of increase as a function of δn calculated here shall be an essential guide in forthcoming experimental investigations of the spontaneous emission of light from a RIF—such a scaling will be a signature that the observed effect indeed arises from vacuum fluctuations at the horizon.

4.3.2 Emission in the Laboratory Frame

Finally, we arrive at the main theoretical result of this Thesis: we will compute spectra of emission from the refractive index front (RIF) moving in a dispersive medium as they can be observed from the laboratory frame. This is a new result, which was calculated for the first time in [15]. Here, we will take more space to comment further on the spectra of [15]—in particular we will study the characteristics of the emission peak in the negative-norm optical-mode, *noL*, which will be the subject of the experimental efforts presented in the last chapter of this dissertation.

The spectra calculated in Sect. 4.3.1 would be observable in the moving frame. In other words, one would need stationary photon counters in the frame moving (with the RIF, at velocity $u = 0.66c$) in the medium to observe the flux density in various modes. Of course, in an actual experiment, the detectors are located on an optical table in the laboratory, and are thus at rest in the laboratory frame. And thus, the spectrum observed is different from those of Sect. 4.3.1. In [14], Finazzi and Carusotto calculated a laboratory frame spectrum for only one mode, which is not the strongest mode. However, this spectrum is not observable for the following reasons: first the refractive index change δn they consider cannot be reached experimentally; second, on either side of the RIF, each moving frame frequency ω' corresponds to up to eight different laboratory frequencies for the 8 modes involved, as in Fig. 3.3. In this section, we will make use of the analytical method we created in Sect. 4.2 to calculate emission from 200 to 7000 nm in the laboratory frame. As we explained when detailing the algorithm, we calculate emission with only positive laboratory group velocity, and we found that emission at a fixed laboratory frequency may arise from several optical modes.

4.3.2.1 Laboratory Frame Spectral Density in Fused Silica

Figure 4.10 shows laboratory spectra in bulk fused silica for three index changes in the low refractive index change regime ($\delta n < 0.04$). As the spectrum is composed of contributions from different modes for different mode configurations, it exhibits a number of sharp features that we will now proceed to describe. Note that we choose to limit the range of optical wavelengths such that no modes in the top branch of the dispersion relation (3.59) are excited, resulting in a cut-off at 230 nm (this will of course be material dependent). Starting from this cut-off wavelength, we first encounter a peak around 250 nm—which is actually the largest spectral density obtained, and corresponds to emission from the negative-norm mode noL. We shall come back to this peak and comment on its features later, but we can already dwell upon its existence: emission is generated by the pairwise coupling of two modes of opposite norm. Mode noL is the only negative-norm mode on the optical branch, and because of the shape of the dispersion in the UV, it covers a rather small laboratory spectral interval (between the solid violet and red vertical lines in Fig. 4.10). Therefore, all emission due to the coupling of two optical modes leaves a contribution within this emission peak in the UV spectral range. The optical emission being by far the strongest, this UV interval contains emitted photons almost every time a photon is emitted at all.

The coupled positive-norm mode (i.e., the partner photon), if optical, can be found at the remaining optical frequencies. Not all coupled mode pairs are separated by a black- or white-hole-type horizon. For example, intervals with horizons, as schematically sketched in Fig. 3.6, are found between the two sets of black and orange dashed lines in Fig. 4.10 but not in the adjacent spectral regions. The short (long) wavelength interval (indicated by arrows in Fig. 4.10) corresponds to a black-hole (white-hole) configuration. We observe that the presence of optical horizons

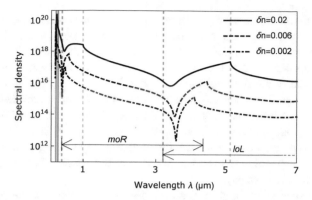

Fig. 4.10 Emission spectral density in the laboratory frame. At each wavelength the total spectral flux density, the number of photons emitted per unit time and unit bandwidth (in photon per nm and per ps), is the sum of contributions from all modes. Emission is concentrated in the UV in a narrow spectral peak generated from mode *noL*. Emission is also strong over spectral horizon-type intervals. Spectra are calculated for wavelengths above the violet line, beyond which there are no contributions from the top dispersion branch. The red line corresponds to $\omega' = 0$ (phase velocity horizon). The black and orange dashed lines indicate the interval of the black-hole (white-hole) mode configuration for the *moR* (*loL*) mode at short (long) free-space wavelengths

leads to an enhancement of the emission. Modes *moR* and *loL* exhibit clear horizon emission profiles[2] between the black and orange dashed lines, and their intervals of emission are indicated by arrows. Over the visible range, emission from *moR* dominates.

Figure 4.10 also shows traces for lower nonlinearities. As expected, the spectral density decreases, and the intervals of optical horizons, associated with strong emission, narrow. The red line at 286 nm corresponds to zero moving-frame frequency ($\omega' = 0$, phase velocity horizon condition); no major spectral features seem to be associated with this position. This is more clearly seen when taking a closer look at the emission peak in the UV.

4.3.2.2   Emission Peak into the Negative-Norm Optical-Mode

We now focus on this peak in the UV: in Fig. 4.11, we plot an excerpt of Fig. 4.10 that shows the interval over which the highest density of emission can be observed. For wavelengths between 230 and 286 nm, emission is due to contributions from light in mode *noL* only. For longer wavelengths (beyond the red line), only mode *uoL* contributes to the emission. We see very clearly that the transition from emission from *noL* to *uoL* is smooth and that no spectral feature is associated with this position.

---

[2]In Sect. 4.3.1 we saw that emission over the (white- and black-hole) horizon intervals is characterised by a "shark fin" shape. We identified this as a signature of horizon physics, as illustrated in Figs. 4.5 and 4.7.

So it seems that the existence of a phase velocity horizon at this wavelength does not influence the rate of emission. Note that this wavelength lies outside of the horizon-type intervals, that are delimited by dotted orange and black vertical lines in Fig. 4.11. Evidently, emission is increased over these horizon-type intervals, and the spectral shapes observed in the moving frame (refer to Fig. 4.11) feature in the laboratory frame spectral density—they are merely mirrored by the effect of the boost: in the moving frame, mode noL has negative group velocity, whilst in the laboratory frame it has positive group velocity. As a result, the UV peak is highly structured in intervals with horizon-type and no horizon emission, with the largest spectral density obtained at 251 nm (corresponding to the onset of the frequency-interval over which the RIF acts as a black hole horizon).

As can be seen from the dispersion relation in the moving frame (see Fig. 3.5), and from the spectra computed in Fig. 4.10, as the height of the RIF (the change in refractive index δn) decreases, the intervals with horizon emission narrow. So the peak at 251 nm in Fig. 4.11 would move to shorter and shorter wavelengths as δn is lowered, and its spectral density would decrease. Referring to the experimental data of [16], in which a fundamental soliton of height $\delta n = 8 \times 10^{-7}$ was propagated in an optical fibre, we can compute the spectrum of light spontaneously emitted from the vacuum of a RIF of height of the order of 10^{-6} in bulk fused silica. This is shown in Fig. 4.12: the high spectral density feature is now extremely narrow, with a bandwidth of 1.5×10^{-4} nm, and a strength of about 2×10^{13} photons per unit time

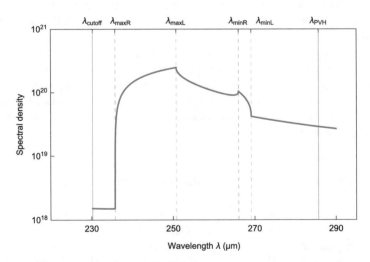

Fig. 4.11 UV peak of the laboratory frame spectral density of emission from a RIF. The RIF (of height $\delta n = 0.02$) in fused silica moves at velocity $u = 0.66c$, and this temporally varying medium excites photon pairs out of the vacuum. For each photon pair created on the optical branch of the dispersive medium, one photon will lie in the interval [230 nm, 286 nm] because that is the interval over which the unique optical mode with negative norm mode contributes to emission as measured in the laboratory frame. The spectrum is structured in intervals of emission with black-hole, white-hole and no horizons

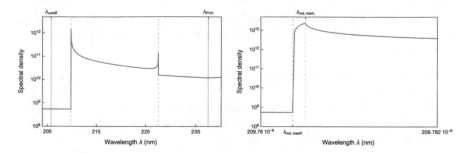

Fig. 4.12 Laboratory frame emission for $\delta n = 2 \times 10^{-6}$ from a RIF moving at $u = 0.66c$ in bulk fused silica. **a** Only the negative-norm optical-optical frequency mode *noL* contributes to the emission between the purple and red lines, and only the positive-norm optical-frequency mode *uoL* contributes to the emission beyond the red line. In **b**, we have zoomed in around the 209 nm region to display the clear black-hole-horizon-type emission feature of the UV peak

and unit bandwidth. The peak remains significantly strong with respect to emission at other wavelengths, which should make it an adequate target for an experiment.

The quantum state at the output is expected to be a two-mode squeezed vacuum state if only two modes were involved (see the argument of Sect. 3.1.2, and in particular the derivation leading to Eq. 3.46). However, the present study makes clear (via Fig. 4.7d for example, or Eq. 3.93) that, for each moving-frame frequency, each mode can couple to up to five positive-norm and three negative-norm modes. Thus we expect the final quantum state on the optical branch to be in a mixed state across the optical modes. Yet, coupling between particular mode pairs seems to dominate in parts of the spectrum—over intervals in which there are horizons—, in particular within the optical branch (see Fig. 4.6d and the almost-equal flux density in *moR* and *noL*). Further characterization of the exact state emerging is needed—and one might want to compute a correlation map between the modes in the moving and in the laboratory frame to scrutinise it.

4.3.3   Conclusion and Discussion

It would be interesting to compute such a spectrum as Fig. 4.11 for a different geometry of the RIF, for example a more realistic pulse shape such as a hyperbolic secant squared, to assess which of the numerous features displayed in Fig. 4.11 are conserved. Efforts in this direction have been pursued in recent publications, see for example [9, 17, 18], in which analytical or numerical calculations in the moving frame were carried for smooth RIF geometries—but no spectrum in the laboratory frame, or even pulse-like geometries (with asymptotically-flat, low-refractive-index, regions on either side of a symmetric bell-shaped RIF) were considered.

In the search for spontaneous emission from an optical setup, the study of the simple geometry of the step-like RIF has proven extremely informative in that it allowed for:

1. clearly establishing the matching conditions between the fields on either side of the interface;
2. clearly identifying the various contributors to emission in positive and negative norm modes—for example, Figs. 4.6d and 4.7d make clear which mode takes part in the quasi-pair-wise emission;
3. creating an analytical method and algorithms to describe the scattering of an incoming field into outgoing fields without approximations—indeed, in contrast with the above-mentioned publications, we consider exact solutions to the fields equations in asymptotically flat regions around the interface, and do not resort to the JWKB approximation, for example;
4. discovering signature features of event horizon physics in dispersive media, such as the increase in the photon flux and the shape of this increase;
5. computing the first spectrum of light spontaneously emitted from the vacuum as it can be observed in the laboratory frame.

Such an analytical method as that presented in the first part of this chapter (Sect. 4.1) can be used to parse a pulse into discrete regions, and to thus calculate a scattering matrix for incoming fields on the right and on the left of the (now spatially symmetric) RIF into outgoing fields (on the right and on the left of the RIF). This original idea of König's has not yielded any result yet, but is under investigation by others in the Quantum Optics group at St Andrews.

In [19], the authors make use of the algorithm developed in [10] to calculate spectra for a smooth hyperbolic secant squared profile in the refractive index of a fibre. Their findings and methods will be compared to these of this Thesis in a later paragraph—see Sect. 5.5.

For the sake of this Thesis, we content ourselves with a step-like RIF geometry and the spectra computed in this chapter.[3] The most interesting feature of these spectra, for what follows, is certainly the high spectral density UV peak. Indeed, as we discussed above, this peak stems from contributions of light in the unique negative-norm optical-frequency mode in the medium. In the mechanism of photon pair creation from the vacuum at an interface, one of the two peers will have a negative norm (like the Hawking partner does in the theory of Chap. 3), so any positive norm photon emitted at a wavelength beyond the UV peak will be correlated to a photon within the peak interval. In particular, photons in this UV peak (more precisely those within the black-hole horizon feature of the peak) will be entangled with photons in

---

[3]As was discussed in the introduction of this chapter, the dispersion relation used to compute the spectra is not that of the medium used in the experiment of Chap. 5: it is that of bulk fused silica (BFS) and differs from that of the photonic crystal fibre (PCF). Considering BFS allowed for checking our results against the literature. In contrast, it is not possible to use a physically meaningful analytical relation for the dispersion of the PCF. This examination is the subject of ongoing work at St Andrews.

the unique mode that allows for light to escape from the boundary (like the Hawking radiation, again).

In an experiment, one cannot realise a step-like but a smooth RIF. Nevertheless, because of its pair-wise emission origin and of the refractive index of materials in the UV, the negative-norm optical peak will always have a large spectral density and narrow bandwidth relative to the rest of the spectrum of spontaneous emission. For most materials that can be used in an optics experiment nowadays (e.g. bulk fused silica, diamond or fused silica PCFs), this peak will lie in the UV. It thus appears as an observable of choice in any experiment that would aim at detecting photons emitted by an optical black-hole horizon.

In the next chapter of this dissertation, we will present the experimental efforts that were conducted in this direction: we will, in particular, see how positive-norm light, in a non-vacuum state, incoming on the RIF would scatter into the negative-norm mode, yielding parametric amplification of the emission in this mode.

References

1. T. Jacobson, Black-hole evaporation and ultrashort distances. Phys. Rev. D **44**, 1731–1739 (1991)
2. W.G. Unruh, Sonic analogue of black holes and the effects of high frequencies on black hole evaporation. Phys. Rev. D **51**, 2827–2838 (1995)
3. I. Carusotto, S. Fagnocchi, A. Recati, R. Balbinot, A. Fabbri, Numerical observation of hawking radiation from acoustic black holes in atomic bose einstein condensates. New J. Phys. **10**(10), 103001 (2008)
4. S. Corley, Computing the spectrum of black hole radiation in the presence of high frequency dispersion: an analytical approach. Phys. Rev. D **57**, 6280–6291 (1998)
5. U. Leonhardt, S. Robertson, Analytical theory of hawking radiation in dispersive media. New J. Phys. **14**(5), 053003 (2012)
6. J. Macher, R. Parentani, Black/white hole radiation from dispersive theories. Phys. Rev. D **79**(12) (2009)
7. S. Corley, T. Jacobson, Hawking spectrum and high frequency dispersion. Phys. Rev. D **54**, 1568–1586 (1996)
8. S. Robertson, Hawking radiation in dispersive media. Ph.D. thesis, University of St Andrews, St Andrews, May 2011
9. S. Robertson, Integral method for the calculation of hawking radiation in dispersive media. II. Asymmetric asymptotics. Phys. Rev. E **90**(5) (2014)
10. S. Robertson, U. Leonhardt, Integral method for the calculation of hawking radiation in dispersive media. I. Symmetric asymptotics. Phys. Rev. E **90**(5) (2014)
11. S. Corley, Particle creation via high frequency dispersion. Phys. Rev. D **55**, 6155–6161 (1997)
12. A. Recati, N. Pavloff, I. Carusotto, Bogoliubov theory of acoustic hawking radiation in bose-einstein condensates. Phys. Rev. A **80**, 043603 (2009)
13. S. Finazzi, R. Parentani, Hawking radiation in dispersive theories, the two regimes. Phys. Rev. D **85**(12) (2012)
14. S. Finazzi, I. Carusotto, Quantum vacuum emission in a nonlinear optical medium illuminated by a strong laser pulse. Phys. Rev. A **87**(2) (2013)
15. M. Jacquet, F. König, Quantum vacuum emission from a refractive-index front. Phys. Rev. A **92**(2) (2015)
16. T.G. Philbin, C. Kuklewicz, S. Robertson, S. Hill, F. König, U. Leonhardt, Fiber-optical analog of the event horizon. Science **319**(5868), 1367–1370 (2008)

17. F. Belgiorno, S.L. Cacciatori, F.D. Piazza, Hawking effect in dielectric media and the hopfield model. Phys. Rev. D **91**(12) (2015)
18. M.F. Linder, R. Schützhold, W.G. Unruh, Derivation of hawking radiation in dispersive dielectric media. Phys. Rev. D **93**(10) (2016)
19. D. Bermudez, U. Leonhardt, Hawking spectrum for a fiber-optical analog of the event horizon. Phys. Rev. A **93**(5) (2016)

Chapter 5
Experimental Observation of Scattering at a Moving RIF

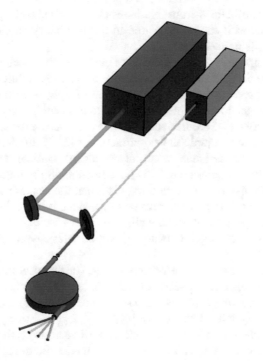

5.1 Stimulated Scattering

Looking back to the previous chapters of this dissertation, we see that we have successfully explained how an analogue to the horizon of a black-hole could be created
by means of a light-induced disturbance in the refractive index of a dispersive dielectric (see Sect. 2.3); and how the study of the conditions under which this happens
can shed light on various aspects of black-hole physics. In particular, in Sect. 3.2,

© Springer International Publishing AG, part of Springer Nature 2018   129
M. J. Jacquet, *Negative Frequency at the Horizon*, Springer Theses,
https://doi.org/10.1007/978-3-319-91071-0_5

we expanded and used a quantum theory for light in a dispersive medium to calculate the rate of spontaneous emission from the vacuum at a moving front in the refractive index (RIF). This study shed some new light on the effect of particle pair creation at the (analogue) horizon: on the one hand, we showed that spontaneous emission—that results from the mixing of a variety of positive and negative norm modes at the RIF—takes place at all frequencies, even at frequencies at which the conditions for the RIF to act as an analogue horizon are not met. On the other hand, we found that emission was stronger around the horizon-like frequency-intervals with a characteristic spectral shape.

Furthermore, over the horizon intervals, quasi pairwise particle production seems to dominate—with strong emission in a unique (positive norm, optical) mode allowing for light to "escape from the horizon" and a negative norm (optical) mode in which light "falls behind the horizon", in a process *à la Hawking*. Interestingly, we also found that contributions from this latter mode (called *noL*) yielded the highest spectral density as it can be measured in the laboratory, in particular over the analogue-black-hole frequency interval.

The findings summarised above stem from the study of the scattering of incoming field modes in the vacuum state, and help understand the conditions that an experiment aimed at observing spontaneous emission of light from the vacuum would have to meet. To date, no such optical experiment has been successfully conducted, but the classical effects of the horizon on waves has been extensively studied and demonstrated—see for example the experiments of [1, 2]. In these experiments, a mode of light in a coherent state was sent on the RIF and the resulting reflection and frequency shift were observed. This phenomenon, in the framework described in Sect. 3.2 of this dissertation, is nothing else than the scattering of an incoming mode of positive norm (populated with photons, since it is a coherent state) in an outgoing mode, of positive norm as well, that allows for light to propagate away from the RIF into the analogue-outside region (the low refractive index region, for $\zeta > 0$, of Fig. 3.4).

Let us cast this statement in a derivation: we can use the scattering theory outlined in Sect. 3.2.4 to calculate the expectation value of *out* modes when an *in* mode is in a coherent state. Denoting α ($\bar{\alpha}$) as a mode of same (opposite) sign in norm as α_0, the incoming state is defined as $\left|\rho_{\alpha_0}\right\rangle = \left|\eta_{\alpha_0}\right\rangle \otimes \sum_{\alpha \neq \alpha_0} \left|0_\alpha\right\rangle \otimes \sum_{\bar{\alpha}} \left|0_{\bar{\alpha}}\right\rangle$. This state is populated with photons in a unique mode of positive norm α_0 that is in a coherent state, whilst all other positive and negative norm modes are in the vacuum vacuum state.

The number operator in an *out* mode α_1 is given by Eq. (3.90). Whence the expectation value for the number of photons in the *out* mode is $\left\langle \hat{N}^{\alpha_1} \right\rangle = \left\langle \eta_{\alpha_0} \left| \hat{N}^{\alpha_1} \right| \eta_{\alpha_0} \right\rangle$. For clarity, we will perform the algebra term by term, beginning with the first term: by $\hat{a}^{\alpha_0} \left|\rho_{\alpha_0}\right\rangle = \eta \left|\eta_0\right\rangle$,

$$\left\langle \rho_{\alpha_0}\right| \sum_{\alpha\alpha'} \beta^{\alpha\alpha_1\star} \beta^{\alpha'\alpha_1} \hat{a}^{\alpha\dagger} \hat{a}^{\alpha'} \left|\rho_{\alpha_0}\right\rangle = \sum_{\alpha\alpha'} \beta^{\alpha\alpha_1\star} \beta^{\alpha'\alpha_1} |\eta|^2 \, \delta_{\alpha'\alpha_0} \delta_{\alpha\alpha_0}$$
$$= |\beta^{\alpha_0\alpha_1}|^2 \, |\eta|^2 \,. \tag{5.1}$$

The second term was calculated in Sect. 3.2.4 by Eq. (3.91) and yields $\sum_{\bar{\alpha}} |\beta^{\bar{\alpha}\alpha_1}|^2$. Now, calling on $\hat{a}^{\bar{X}}\hat{a}^{\bar{X}\dagger} |\eta_X\rangle = \left\langle \eta^{\bar{X}} \right\rangle |\eta_X\rangle + |\eta_X\rangle$, and recalling that the incoming modes of negative norm are in the vacuum state, $|0_{\bar{\alpha}}\rangle$, which implies that the first term goes to zero, it follows that $\hat{a}^{\bar{\alpha}\dagger}\hat{a}^{\bar{X}}\hat{a}^{\bar{\alpha}\dagger} = |1\rangle_X$. Wherefrom the third term of (3.90) goes to zero:

$$\langle\rho_{\alpha_0}| \sum_{\alpha\bar{\alpha}'} \beta^{\alpha\alpha_1\star}\beta^{\bar{\alpha}'\alpha_1\star}\hat{a}^{\alpha\dagger}\hat{a}^{\bar{\alpha}'\dagger} |\rho_{\alpha_0}\rangle = \eta^{\star}\delta_{\alpha_0\alpha_1}\left\langle \eta_{\alpha_0} |1_{\bar{\alpha}'} \right\rangle = 0, \qquad (5.2)$$

and so does the fourth one-

$$\langle\rho_{\alpha_0}| \sum_{\bar{\alpha}\alpha'} \beta^{\bar{\alpha}\alpha_1}\beta^{\alpha'\alpha_1}\hat{a}^{\bar{\alpha}}\hat{a}^{\alpha'} |\rho_{\alpha_0}\rangle = \langle 1_{\bar{\alpha}}| \delta_{\alpha_0\alpha_1}\eta |\eta_{\alpha_0}\rangle = 0. \qquad (5.3)$$

Therefore, the expectation value in an *out* mode when one of the positive-norm *in* modes is in a coherent state is

$$\left\langle \hat{N}^{\alpha_1} \right\rangle = |\beta^{\alpha_0\alpha_1}|^2|\eta|^2 + \sum_{\bar{\alpha}} |\beta^{\bar{\alpha}\alpha_1}|^2. \qquad (5.4)$$

The second term is the expected contribution from the vacuum (an incoming mode of opposite sign in norm to the considered out mode will scatter into it and the action of its creation operator on the vacuum will result in a photon being emitted). The first term is due to, in quantum optics, parametric amplification: an in state comprising photons in one positive norm mode only will stimulate emission into any out state, irrespective of the sign of the norm of the latter. Indeed, upon introducing the *out* mode α_1, we have made no assumption regarding the sign of its norm—the above derivation actually holds for both positive- and negative-norm *out* modes. To the best of my knowledge, this is an original finding of this Thesis: it is the first time that the stimulation of outgoing states of negative norm by monochromatic coherent input states of positive norm is introduced in the field of optical analogues.

Two experiments performed by Rousseaux et al. [3] and Weinfurtner et al. [4] already demonstrated the phenomenon of stimulated scattering from a positive to a negative norm wave at the group velocity horizon in a water-based setup. Note that there is a debate in the community as to whether these water wave experiments were performed in the linear or nonlinear regime [5, 6]. In that regard, the 2016 experiment by Rousseaux and collaborators [7] more clearly demonstrates physics belonging to the linear regime. Likewise, the first experimental observation of negative-norm light in optics, by König, Faccio and collaborators in 2012, was performed in the nonlinear regime [8]. In that experiment (see Sect. 2.2.2 for a more detailed discussion), the pulse sent in the dielectrics was well above the power-level for fundamental solitons. Thus it is a nonlinear effect, whereby photons *from the pulse* were scattered to different frequencies—including the Negative Resonant Radiation (NRR) in the

UV—that was observed.[1] Although the observation [8] was a convincing evidence of negative-norm light created by parametric generation, this experiment did not provide well-defined input mode frequencies as the stimulation resulted from the broad spectrum of a collapsing pulse. Furthermore, because of the high intensity, and the induced nonlinear propagation of the pulse, it is unclear whether the scattering of the incoming modes (from the pulse itself) into outgoing modes can be described by the (linear) physics of event horizons.

In terms of the optics experiments [1, 2], this means that the observed (positive norm) frequency shifts were accompanied by transfer of energy from the probe to a wave of negative norm. This effect was not observed at the time, though, and we here propose to do so. For this purpose, we assemble an experiment allowing for an incoming laser beam to impinge on a fundamental soliton propagating in an optical fibre, to probe the effect of energy transfer to a negative norm wave at the analogue horizon. This chapter is the experimental component of the Thesis, and shall culminate in comments on the observation of the above-mentioned effect in fibre optics. In the next section, we will present the setup assembled to preform the experiment. We will explain the choice of the laser used to create the soliton in the optical fibre, and that of the frequency filtering and measurement techniques used to observe frequency shifting at the horizon. We will then dedicate a section to the observation of positive-norm to positive-norm frequency shifts, and comment on the various theories that can describe this effect. Following on which, the chapter will conclude with the presentation of the efforts made towards the detection of light scattered from the positive-norm *in* mode to the negative-norm *out* mode—that lies in the UV in the present case. Finally, we shall make some remarks on the present endeavour and look out to the future of this experiment.

5.2   A Journey, at the Speed of Light, on an Optical Table

In this section of the dissertation, we will look at all the apparatus and optical elements used in the setup that was created to observe the frequency shifting of light at the optical horizon. The optical elements fall into two categories: the optics used to prepare the state of light sent in the fibre, and the optics used to collect the output and distribute it to the various measurement devices. Likewise, the apparatus can be separated between the lasers used to create the input, and the measurement devices we chose to probe the output. All are arranged around the optical fibre in which the phenomena of interest take place. This section presents all of these in the order

---

[1]The experiment that was realised for this Thesis relies on the propagation of a fundamental soliton in the fibre. This modifies the refractive index by the Kerr effect—which is a nonlinear effect—and the photons that are scattered at the pulse edges come from a different light field, a probe. The latter scatters *linearly* on the refractive index front: although this front is created by nonlinear means, the scattering described by Eq. (5.4) and implemented in the experiment is a linear process, unlike the generation of NRR from higher-order soliton pulses.

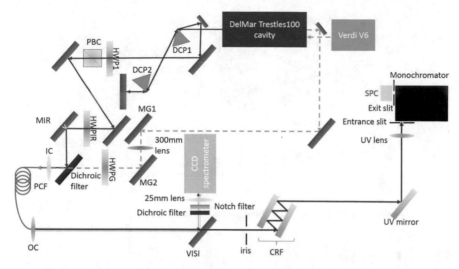

Fig. 5.1 Schematic of the experimental setup. The Verdi V6 laser delivers a single frequency CW beam at 532 nm that is used to pump the DelMar cavity and coupled in the fibre (PCF) and used as a probe wave. The DelMar cavity outputs ultrashort, pJ-energy, pulses of central wavelength 750 to 890 nm that are propagated through a pair of dispersion compensation prisms (DCP1 and DCP2) twice to create negative group velocity dispersion (GVD) and compensate for the broadening of the pulses upon interaction with all other optics. HWP1 and PBC are used to control the amount of IR light that reaches the tip of the fibre. HWPIR and HWPG are used to align the polarization of the pulse that generates the fundamental soliton in the fibre and that of the probe wave. Visible-IR reflectors, MG1 and MG2 for the green, MIR and the dichroic filter for the IR, are used to aim the co-polarised input beams through the input coupler IC on the tip of the PCF. The UV component in the output beam is collimated by the output coupler OC (UV-condenser triplet). The spectral characteristics of the visible and IR components of the light can be measured by inserting VISI in the output beam, and further inserting a dichroic and notch filters before the CCD spectrometer. The UV component of the beam is isolated from all other light by spatially and frequency filtering the output beam by means of iris and the cascaded reflection filter CRF (composed of 2 UV mirrors) and UV mirrors, respectively. UV light is then probed with sub nm precision by the single photon counter SPC installed behind the monochromator

one would encounter them in when journeying on the optical table, from the lasers through the fibre and to the measurement devices—as shown on the bird-eye view schematic of the setup in Fig. 5.1.

5.2.1 Pulse and Probe Sources

In the experiment, the front in the refractive index (RIF) of the fibre is realised by means of an intense few-cycle pulse that creates a soliton (see Sect. 2.2.1 for the theoretical description of this effect). The RIF is probed with a continuous wave (CW) laser beam of wavelength 532 nm.

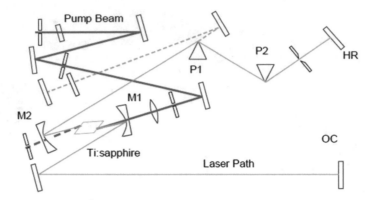

Fig. 5.2   Schematic of the Trestles 100 folded cavity, DelMar Photonics

5.2.1.1   Probe Laser

The probe wave is provided by a CW laser at 532 nm wavelength (Verdi V6, Coherent Inc.), that delivers 6 W of output power in the TEM_{00} transverse mode. The output spectrum of this laser is actually an extremely narrow line at 532 nm, of width less than 5 MHz.[2] This single-frequency (single longitudinal mode) operation makes it ideal to probe the effect of frequency-shifting at the RIF: the coherent state that will be scattered at the RIF will be of a single, monochromatic, comoving frequency ω', as in the calculations (5.4).

5.2.1.2   Pulse Laser

A 3W, 532 nm beam provided by the Verdi laser is coupled to the cavity of a Titanium:Sapphire laser (Trestles100, Del Mar Photonics, Inc.) and used to pump it to create pulses shorter than 100 fs, with a repetition rate of 81 MHz, in the IR.

This laser is based on a folded cavity design (see Fig. 5.2, composed of 10 mirrors (including M1, M2, HR and the output coupler (OC)), a Titanium-Sapphire crystal (TiS), a lens for the focusing of the pump radiation (L), two prisms (P1 and P2), and a slit (S). This folded design results, in particular, in a virtually astigmatism-free output beam. It also enables the tuning of the central wavelength of the IR pulses—in the experiments described in this thesis, I could create pulses of central wavelength, λ_c, 749–887 nm (the Ti:Sapphire gain range extends from 710 to 950 nm) of duration between 50 and 85 fs. These pulses are created by means of the Kerr lens mode-locking (KLM) principle of operation [9].

As we will see later, the experiment solely requires the propagation of a fundamental soliton in the fibre, thus the IR power impinging on the tip of the fibre ought

---

[2]Manufacturer data, measured over 50ms with a thermally stabilized reference etalon at maximum specified output power.

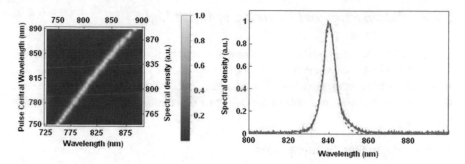

Fig. 5.3 Output spectra of the pulse laser (Trestles100, Delmar Photonics Inc.). The central wavelength of the pulses is tuned from 750 to 890 nm. The (normalised) spectral density of the pulse at all wavelengths is characteristic of a cavity soliton—it can be fitted with hyperbolic secant shape, as exemplified for the $\lambda_c = 840$ nm pulse ($\Delta\lambda = 8.8$ nm, $T_0 = 48$ fs, $E_P = 8.94$ pJ)

to be much lower than the maximal output power attainable. I obtained average powers above 200 mW for pulses of central wavelength λ_c between 790 and 870 nm. For output pulses of central wavelength beyond this range, I obtained lower average powers, typically below 190 mW. I observed that the pulses at the output of the laser all had a spectral density that could be fitted with a hyperbolic secant function, as in Fig. 5.3. This is a signature of cavity solitons, i.e., the pulses propagating inside the cavity are discrete solitons.

5.2.1.3   Keeping the Pulses Short

As can be seen in Fig. 5.3, the pulse emitted by the pulse laser are relatively broadband, with a bandwidth (measured at the full width at half maximum) $\Delta\lambda = 8.8$ nm at 840 nm, corresponding, by the time-bandwidth product $\Delta\lambda\Delta t \geq 0.315\lambda_0^2/c$ [9], to a pulse length $\tau_{FWHM} = 84$fs. Upon reflection on surfaces, or transmission through the material of the various optical elements on the way to the tip of the optical fibre, such short-broadband pulses experience positive group velocity dispersion (GVD) and thus broaden. To compensate for this broadening, which is detrimental to the control of the soliton ultimately generated in the fibre, the beam is passed twice through a pair of dispersion compensation prisms (DCP1 and DCP2) by means of a reflection on a mirror positioned after the second prism (as can be seen on Fig. 5.1). Careful alignment of the beam path through the pair allows to create negative (or anomalous) dispersion: high frequency components are made to travel faster than the lower ones, thus negatively chirping the pulse (see the Ph.D. dissertation of McLenhagan [10] for the various GVD and GDD techniques developed by the St Andrews group). Two passages through the pair result in a negative GVD that compensate for the interaction with other optical elements further in the beam path down to the fibre tip, and ensures that the pulse used to create the soliton is indeed an unchirped hyperbolic secant one.

5.2.2   *Polarisation and Coupling of Input Light*

Continuing our journey on the optical table towards the tip of the fibre, alongside the IR pulse, we first encounter some polarisation optics. As was mentioned earlier, and will be calculated later, the realisation of a RIF in the optical fibre does not require the full average power available at the output of the Trestles100.

5.2.2.1   Polarising the Input Light

In order to control the power impinging on the tip of the fibre, we use a combination of polarisation optics, a half-wave plate (WPH05M 808, Thorlabs Inc.) and a polarising beam cube (PBS052, Thorlabs Inc.), labelled as HWP1 and PBC, respectively, in Fig. 5.1. Both intended for use over the spectral range 620–1000 nm, which, as can be seen on Fig. 5.3, covers the full pulse bandwidth for all central wavelengths. The IR beam is linearly polarised, along an axis that can be rotated, by HWP1. As a function of the polarisation state sent on the PBC, none, a fraction, or all of the power (up to Fresnel reflection) will be transmitted through it. Behind the PBC, and before the IR beam encounters the coupling optics, it goes through a second half-wave plate, HWPIR (WPH05M 808, Thorlabs Inc.) on Fig. 5.1. HWPIR is used to rotate the linear polarisation of the IR beam and to align it to either the polarisation axes of the fibre, or in parallel with the green beam.

The green probe consists of the, undepleted, remains of the 3 W of 532 nm light used to pump the pulse laser: mirrors M1 and M2 (that constitute the telescope around the Ti:Sapphire crystal in the Trestles100 cavity) are transparent at 532 nm, and let 300 mW of the pump beam escape the cavity via an aperture (as can be seen on Fig. 5.2). This Gaussian beam is directed towards the coupling optics by reflection on 4 mirrors—in particular MG1 and MG2 (see Fig. 5.1) are used to aim the beam through the coupling lens onto the tip of the fibre. After MG2, a half-wave plate (HWPG, WPH05M-532) specified for 532 nm is used to rotate the linear polarisation of the green beam. In the experiment, the input IR and green beams are co-polarised. The two half-wave plates, HWPG and HWPIR, are not exactly aligned in their respective mounts, as can be seen on Fig. 5.4. This was gauged by measuring the power in the light reflected on a glass slide set at the Brewster angle in the linearly polarised beam as is diagrammatically illustrated in Fig. 5.4: a microscope slide is placed at 104° in the IR (green) beam reflected (transmitted) through a dichroic filter—the role of which will be explained later—and the power meter head collects light reflected (in a p-polarisation state) from the slide. The amount of light in this reflected beam varies as HWPIR, or HWPG, is rotated, and the result is shown in Fig. 5.4. We see that there is a difference of 60° between the, e.g., horizontal axis of the two waveplates.

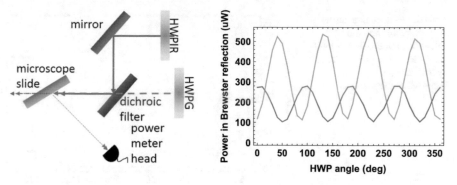

Fig. 5.4 Measurement of half-wave plates axes alignment by Brewster reflection. The incoming polarisation of the IR and green beam is controlled via rotation of HWPIR and HWPG, respectively. P-polarised light (after reflection on, for the IR, or transmission through, for the green, a dichroic filter) is reflected on the microscope slide (inserted at 104°, the Brewster angle, in the beams), and the resulting power is collected on the power meter head. The power in the Brewster reflection is a Sin function of the HWP angle

5.2.2.2 Coupling Light in the Fibre

Input light, from the IR and green beams, has to be focused on the tip of the fibre, to a waist of less than 2 μm (to match the surface area of the core of the fibre, which dimensions will be detailed later) along the direction of the fibre axis. This implies that the two beams have to be made collinear and then focused by the same optical element. To achieve collinearity, the IR light is reflected onto the coupling optics by means of a dichroic filter (660IK25, Comar Optics Ltd), whilst the green beam is transmitted through it towards the fibre. As can be seen from the manufacturer data in Fig. 5.5, the dichroic filter has a high reflectivity, of almost 100%, over the wavelength range of the IR laser beam, and a transmission of close to 95 % at 532 nm—at 45° incidence.

The collinear IR and green beam are then focused on the tip of the fibre by a ($f = 3.1$ mm) aspheric lens (C330TMD-A, Thorlabs Inc.), whose coating reflectivity

Fig. 5.5 Transmission curves (at normal and 45 degree incidence) of the dichroic filter (660IK25, Comar Optics Ltd)—Manufacturer data

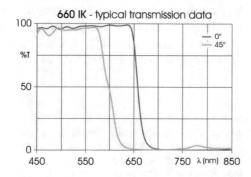

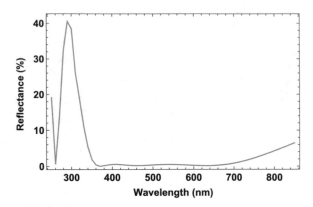

Fig. 5.6 Reflectance curve of type "A" broadband anti-reflection coating of the C330TMD-A aspheric lens by Thorlabs, Inc.—Manufacturer data

is plotted in Fig. 5.6. This lens transmits close to a 100 % of the light in the IR and green beam, in particular, at 532 nm, the reflectance is of only 0.5%, making it suitable to focus the green beam onto the tip of the fibre.[3] Unfortunately, this lens suffers from chromatic aberrations: the green and IR foci are not found at the same point along the lens-fibre axis. To compensate this, the green light is focused by a ($f = 300$ mm) lens before the filter, at the point at which light propagated from the back end of the fibre, and back through the coupling lens was focused. In doing so, I was able to couple up to 60 mW of green light into the fibre—for a coupling efficiency of 20%. As a function of the central wavelength of the IR pulse, I could achieve an IR-coupling efficiency of 21 to 39%.

5.2.3 Optical Fibre

Now we arrive at the central element of the setup, the medium in which the pulses generate the RIF that scatters the green probe to other, positive- and negative-norm modes of oscillation of the electromagnetic field. Our theoretical study of Sect. 2.2.1 will now seem less fortuitous: the medium of choice for the optical study of analogue horizons is a photonic crystal fibre (PCF). PCFs are a particular type of optical fibres engineered to efficiently confine high intensity, ultrashort, pulses of short-IR wavelength in their core. In particular, the dispersion of PCFs can be tailored to generate a regime of anomalous dispersion for such short-IR wavelengths as those delivered by the Trestles100 (in the 800 nm regime, see Fig. 5.3).

---

[3]Again, although only 90 to 98 % of the incoming IR is transmitted through the lens, and thus focused on the tip of the fibre, this does not matter in this experiment: indeed, the laser delivers an output power sufficient to accommodate for the losses at the lens, and all the optics on the way to the fibre, and nevertheless generate a fundamental soliton in the fibre.

5.2.3.1   Characteristics of the Fibre

Recalling the Nonlinear Schrödinger equation stated in Sect. 2.2.1, Eq. (2.62), we identified β_2 as the second order dispersion parameter, or GVD parameter, $\beta_2 = \frac{d^2}{d\omega^2} n(\omega)\omega/c$ with $n(\omega)$ the effective (frequency-dependent) refractive index. For bulk fused silica, of which conventional fibres are made, it can be calculated that β_2 is zero at $\lambda = 1270$ nm [10]. We called this wavelength the zero dispersion wavelength (ZDW). The region of anomalous dispersion lies at wavelengths longer than the ZDW. In fibres, there is a waveguide contribution to the dispersion that shifts the ZDW. This contribution can be made large enough by engineering PCFs, to move the region of anomalous dispersion to wavelengths as short as that of the Trestles100, or even the into visible. There exists many PCFs design, and they all have in common the main idea of surrounding the wavelength-size core of the fibre with a pattern of wavelength-size air holes running the length of the PCF. The two main core designs are a "missing" hole, or solid-core, or a hollow core.

The fibre used in this experiment is the NL-1.5-670 (Blaze Photonics), that has a solid core of diameter 1.5 μm surrounded by a microstructure of ≈2 μm holes (the cladding), as can be seen in Fig. 5.8. Its ZDW is 670 nm, thus pulses of central wavelength 790–870 nm propagate in the anomalous dispersion region of the fibre and can thus generate solitons. With the fibre parameters communicated by the manufacturer, a nonlinear coefficient, γ, of 250/W/km and a group velocity dispersion, D, of 145 ps/nm/km (see Fig. 5.7), we can use Eq. (2.64) to calculate the peak power of fundamental solitons generated at all wavelengths on Fig. 5.3: we obtain peak powers on the order of $P_P = 200$ W, see Fig. 5.9a. The average power necessary to generate such solitons is calculated by [9]

$$P_{avg} = \nu_{rep} \, P_P \, \tau_{FWHM}, \tag{5.5}$$

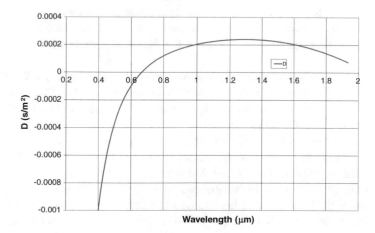

Fig. 5.7 Dispersion curve of NL-1.5-670. Data courtesy of Blaze Photonics

Fig. 5.8 SEM image of the
tip of PCF NL-1.5-670. Data
acquired by Andrea Di
Falco—University of St
Andrews

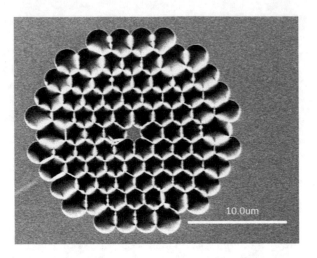

with $\tau_{FWHM} = 1.763T_0$, for a hyperbolic secant pulse, and the pulse duration specified above: 50fs (yielding a T_0 of 28.4fs for a repetition rate of 81 MHz). The average
power depends on the central wavelength λ_0 of the pulse, as in Fig. 5.9b, and is on
the order of 1 mW for an $N = 1$ (fundamental) soliton. As was repeatedly stated in
the preceding argument, the average power delivered by the Trestles100 is in large
excess of what is needed to generate a fundamental soliton in the fibre. Indeed, even
with a coupling efficiency of only 20% (lowest efficiency achieved in this experiment), a beam of only 5mW would suffice. This figure also comes as a confirmation
of the adequacy of the experiment design so far: few optical elements (6 mirrors, two
HWP, a PBC, a dichroic filter and a lens) induce positive GVD in the pulse (for a
total of $\approx +375\text{fs}^2$), that is compensated for by the two passages through the pair of
dispersion compensation prisms and the fibre (that has a GVD of $-41.1\text{fs}^2/\text{cm}$ [10]).
This design enables control over the three input pulse parameters of importance in
this experiment: the input energy, the input polarisation, and the duration (and chirp)
of the pulse upon impinging on the tip of the fibre and generating a fundamental
soliton.

In addition to their short-wavelength anomalous-dispersion region, and their ability to confine high-intensity IR light in their core via mode sieving,[4] PCFs also allow
for long-distance undisturbed propagation of the fundamental soliton by minimis-
ing pulse broadening through propagation—thus allowing for nonlinear interactions
over long distances. And, thus, for long distances soliton-probe interactions—in the
experiments presented in this thesis, the probe scattered on the edge of the soliton
over the total length of the fibre, 1.2 m. The output of the fibre falls in two categories:
strong output (the green and IR components of the beam) and weak output (the UV
component of the beam). These have to be measured with different apparatus and
have to be isolated from each other, as will be detailed in the coming two sections.

---

[4]The lower order modes, that have a large cross section, cannot escape from the core of the fibre
because the "wires" between the cladding holes are too narrow [11].

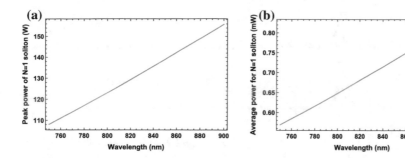

Fig. 5.9 Fundamental ($N = 1$) soliton in the fibre (PCF NL-1.5-670). **a** Peak power of a fundamental soliton generated in the fibre for increasing central wavelength from 750 to 900 nm. The peak power is calculated by Eq. 2.64. **b** Average power necessary to generate a fundamental soliton in the fibre for increasing central wavelength from 750 to 900 nm. The average power is calculated by Eq. (5.5)

5.2.3.2 Fundamental and Higher Order Solitons in the Fibre

The experiments presented in this Thesis rely on the fine control of the order (fundamental, $N = 1$, or higher order) of the soliton that is propagated in the fibre. This order can be determined by comparing the spectrum at the output of the fibre with that of the input pulse for varying output powers. The pulse propagating in the fibre generates a fundamental soliton when the two traces overlap: up to a redshift of the central wavelength due to Raman scattering, the spectrum of the fundamental soliton will be identical in wavelength and energy (surface area in temporal space) to that of the incoming pulse.

I acquired spectra of the IR pulses in the fibre for increasing output power from 0.25 to 16 mW at all wavelengths from 749 to 887 nm. An example of the study of the soliton number is shown on Fig. 5.10 for a input pulse of 840 nm central wavelength (as on Fig. 5.3): the average output power necessary to the formation of a fundamental, $N = 1$, soliton is 1.25 mW. A $\tau = 93$ fs, $\lambda_c = 842$ nm soliton, of bandwidth 8 nm and energy 8.3 pJ then forms in the fibre. Note that this average power is larger than what can be calculated by (5.5) and is shown in Fig. 5.9. This is due to the coupling efficiency at this wavelength: it being low, the mode must overlap with the cladding and only a fraction of its energy is propagated in the core of the PCF, where fundamental soliton are generated and propagate. The spectral shape of the soliton slightly differs from that of the input pulse because of the rotation of the polarisation axes of the fibre over the full length of the fibre (and possibly as a result of the overlap of the input mode with the cladding and core of the PCF).

For higher average output power, the propagation of the IR pulse results in the formation of two pulses that move away from each other in frequency because of Raman scattering. Input pulses that have a shorter wavelength, closer to the ZDW of the fibre will also generate a dispersive wave whose wavelength will shorten as the average power in the fibre is increased.

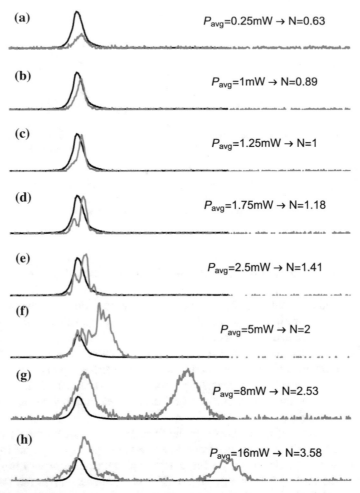

Fig. 5.10 Soliton order as a function of the average output power. An input pulse of central wavelength 840 nm (black) is propagated in the PCF. The input and output spectra are normalised to the $N = 1$ soliton energy. Output spectra for increasing average power P_{avg} from 0.25 to 16 mW are measured, corresponding to soliton numbers of **a** 0.63, **b** 0.89, **c** 1 ($P = 1.25$ mW and $E_{peak} = 8.3$ mJ), **d** 1.18, **e** 1.41, **f** 2, **g** 2.53, and **h** 3.58. Note the change in the scale of the spectra. The spectra are plotted against wavelength, from $\lambda = 800$ nm to $\lambda = 1100$ nm

5.2.4 Strong Output Measurements Setup

Both ends of the fibre are set on two three-dimensional micrometer piezo stages, that allow for further refinement of the tip-(in/out)coupling-optics alignment. Light from the *end* tip of the fibre is collected by a 15 mm UV-condenser triplet (NT 49-693 aspheric UV lens, 0.5 NA, Edmund Optics) chosen for its Numerical aperture (NA) that matches the NA of the fibre in the UV, and its coating and focal length that are specified in the UV—where the signal ultimately lies. Because of the strong chromatic aberration of these optics, "visible" light (at 532 nm and the (red or blue) frequency shifted light) are not collimated when light in the UV is. Because the final aim was to measure an extremely weak (few photons) signal in the UV, I chose to collimate the UV beam and to use a diverging visible beam to measure the positive-norm-to-positive-norm frequency shifts. This choice will be commented further in Sect. 5.4.1.1. Visible light from the fibre is directed to the CCD spectrometer (AvaSpec-ULS2048-EVO, Avantes BV) by a reflector (BB5-EO2, Thorlabs Inc.) inserted in the beam (VISI)—thus simultaneous measurements of the positive-norm-to-positive-norm and positive-norm-to-negative-norm scattering processes is not possible with this setup. This is a drawback only in terms of the duration of the measurement procedure and does not impact the final result. On the contrary, a setup allowing for simultaneous measurements had initially been designed but was found to be too lossy in the UV.

5.2.4.1 Frequency Filtering in the Visible

As will be calculated in the following section of this dissertation (see Sect. 5.3.1), only a certain amount of the probe energy actually scatters off the edges of the pulse. Thus there is a strong beam exiting the fibre at 532 nm wavelength, that has to be filtered out to allow for measuring the (approximate) spectral density in, and wavelength of, the frequency shifted light. The dielectric mirror VISI will reflect over 99% of the incoming light for any polarisation state of the visible light. In order to avoid saturation by the green or IR output in the spectrum regions of the red or blue shifted light, two dichroic filters are inserted in the beam after the mirror: another short-pass filter (660IK25) to filter out the infrared light, and a notch filter centred at 532 nm (NF03-532E-25, Laser2000 Ltd.) to suppress the strong 532 nm component left-over from the probe-pulse interaction. The notch filter has a suppression bandwidth of 17 nm centred around 533 nm, where it effectively acts as a 63 dB attenuator, see Fig. 5.11. Additionally, I found that, contrarily to what Fig. 5.11 suggests, the notch filter actually heavily blocks light in the UV, with an attenuation of over 30dB over the range 220–280 nm. Incoming light is focused on the entrance slit of the CCD spectrometer by means of a 15 mm lens.

5.2.5 Probing the Invisible

As was said earlier, the output light is collected by a 15 mm UV-triplet lens that allows for collimating the UV component of the beam. This UV beam is directed at the UV-measurement apparatus, that consists of an avalanche photomultiplier tube (single photon counter, SPC) placed behind a monochromator. In the experiment, we measure the signal rate in photons per second (Hz) down to the single photon regime. To direct the beam to the measurement apparatus, UV reflectors have to be

Fig. 5.11 Transmittance curve of the notch filter (NF03-532E-25, Laser 2000 Ltd.). Manufacturer data

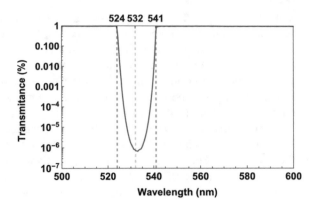

used. Indeed, optics that do not have a coating specifically made for the UV will strongly absorb over this wavelength range—this is, for example, the case with the above-mentioned visible reflector.

5.2.5.1 Improving the SNR by Filtering Out Contaminating Light

As we will see in a following section of this dissertation, we expect a signal of very few photons per second, so each of those that come out of the fibre are a precious and scarce resource that has to be carefully handled on the way to the measurement apparatus. At the same time, the visible and IR components of the output beam are much stronger, by many orders of magnitude: the average IR output power for the fundamental soliton regime is on the order of 1 mW (see Sect. 5.3.3) whilst the output power of the green is of about 60mW—these would create rates of about 4.2×10^{15}Hz and 1.6×10^{17}Hz, respectively, on the SPC. I observed that, if unfiltered, the green and IR components of the output beam would smear out in the monochromator: green and IR photons scatter off the grating and yield background counts at all wavelengths. This "smearing effect" is not well understood and is not linear in wavelength. I measured that the green component of the beam creates a background of 2, 000 Hz, whilst the IR component creates a background of only 100 Hz over the range Thus, the green and IR have to be filtered out of the beam that will eventually reach the monochromator to reach a regime of signal to noise ration (SNR) allowing for detection of a few-Hz UV-signal. In this section, I will discuss the various filtering techniques that were developed and implemented in the experiment to isolate the UV signal from light that contaminates the SNR—the green and IR components of the beam and the background light of the laboratory. These are: the use of UV-coated optics (mirrors and lenses) that absorb long-wavelength (green and IR) light, the insertion of a UV-bandpass filter, and the spatial isolation from the (diverging) long-wavelength components of the beam and background light of the laboratory.

All the optics used in the experiment are standard, off-the-shelf, components that are not well specified in the 220–240 nm wavelength-range (for example, the reflectivity of UV-coated mirrors over this range is not given by the manufacturer)—thus we had to measure these properties. In all the experiments aimed at gauging the UV abilities and characteristics of our setup, we used a DHS lamp (AvaLight-DH-S, Avantes Ltd.), that emits a relatively structured spectrum over the interval 220–260 nm, see Fig. 5.12. We measured the reflectivity of our UV-coated mirrors, "UV mirrors" (11-1620, Optarius Ltd.), in the region 220–240 nm by measuring the power a UV-bandpass filter (228 fs 25–25, Andover Corp.) would let through before the collimated beam from the DHS lamp had been reflected on the mirrors, and after reflection. The reflectivity of the UV mirrors at near-normal incidence was measured to be of about 92%.

A usual isolation technique in optics consists in inserting interference filters in the beam, that only transmit light over a narrow wavelength interval. The UV-bandpass filter is very efficient at filtering out light, even over its transmission bandwidth, as can be seen on Fig. 5.13. For example, the above mentioned background counts detected at all UV wavelengths as a result of the smearing of the strong green and IR components of the beam are totally suppressed by this filter. However, its peak transmittance is only 25% at 228 nm, with a narrow 20 nm bandwidth. Thus, any experimental setup featuring this filter and the UV mirrors would discard so much UV light on the way to the entrance slit of the monochromator that it would not be a sustainable option (further arguments supporting this statement will be given in a following paragraph, see 5.4.1.1). Thus, we could not include it in our setup and had to resort to other filtering techniques to filter the green component out of the beam.

As can be seen in Fig. 5.1, we used an arrangement of three double bounces on a pair of UV mirrors to filter the green and IR components of the beam out. We

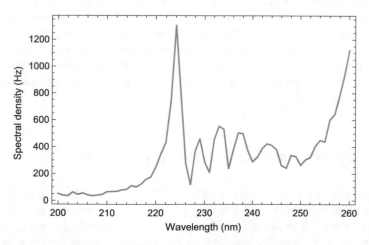

Fig. 5.12 Spectrum of the bare, collimated, beam of the DHS lamp (AvaLight-DH-S, Avantes Ltd.). Data measured with the SPC and monochromator

Fig. 5.13 Transmission
curve of the UV-bandpass
filter (228fs 25-25, Andover
Corp.). This filter has a 20
nm FWHM transmission
bandwidth. Manufacturer
data

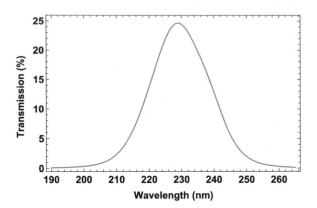

coupled the green laser only in the fibre, and measured an output power of 60mW,
and found that only 2 μW of these remained after the beam had passed through the
"cascaded reflection filter" (CRF). So the beam was attenuated by a factor 3×10^4—
4 orders of magnitude or (40 dB). CRF transmits 62% of the (220–240nm) UV light
provided that the collimated component of the beam has a diameter ≤ 4 mm. Despite
the strong attenuation of the green and IR component of the beams, the smearing
effect still creates a background of the order of 100 Hz and 10 Hz, respectively. In
order to further reduce this background, we installed an aperture ('iris' in Fig. 5.1) to
spatially filter the green (and diverging) component of the fibre-output beam. Further
comments will be made on this when the measurements are presented.

The decontaminated beam is directed toward the entrance slit of the monochromator, on which it is focused by a 100 mm UV lens (PC UV 248–400 nm AR,
Comar Ltd.). The UV detection apparatus consists of a single photon counter (SPC)
positioned behind a Czerny-Turner monochromator. The latter is made of a pair of
curved mirrors arranged around a grating. The first mirror spreads the light on the
grating,[5] and the second one refocuses the diffracted light onto the exit slit of the box,
behind which the single photon counter sits. As we will see shortly, this monochromator (Acton SpectraPro 2500i, Princeton Instruments Inc.) allows for the number
of photons in the beam to be measured for a single wavelength with sub-nanometer
precision by the SPC.

5.2.6 Spectral Sensitivity of the Setup in the UV

In this section, I present the investigation of the spectral sensitivity of the detection apparatus in the UV. In particular, the edge of the spectral sensitivity of the
monochromator and its coupling optics lies close to the wavelength region at which
the signal is expected to be observed. At this edge the spectral response becomes

---

[5]UV holographic grating 1800 g/mm, 360° turret, ARC-1-36HUV, Princeton Instruments Inc.

Fig. 5.14 Quantum efficiency of the PMA series single photon counters. Relevant curve is in red (PMA 182, PicoQuant GmbH). Manufacturer data

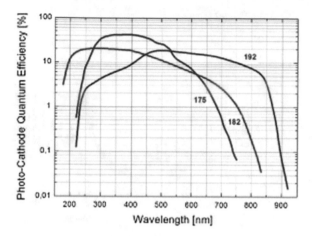

highly non-uniform and therefore a calibration procedure is required.[6] This spectral sensitivity will be uniquely set by the monochromator box: indeed, the single photon counters are based on avalanche photomultiplier tubes (PMA 182, PicoQuant) that have a set quantum efficiency in the relevant UV region of 15 to 20%, see Fig. 5.14. The monochromator, on the other hand, will have a varying sensitivity as a function of the wavelength—this is what we need to measure.[7]

5.2.6.1   Spectral Resolution in the UV

In order to measure the spectral resolution and bandwidth of the monochromators, we used a commercial UV lamp provided to us by the Organic Semiconductors Centre (OSC) at St Andrews in their explosive-detection experiments: the "Tiramisu" lamp—the spectrum of which was measured when they lent it to us and is shown on Fig. 5.15. This spectrum has many features, and we decided to focus on three of these that have a very narrow bandwidth: the falling ramp at 192 nm, and the peaks at 250 and 365 nm. The lack of information on the chemical content of the radiating plasma implies that it was impossible to determine whether these features were characteristic of emission lines or other emission processes, but this did not prevent the study from being carried out because they appeared to be very stable

---

[6]This had already been investigated by Dr McLenaghan for her Thesis (see [10]) but I found that the setup that was designed at the time was based on mirrors that had a low reflectivity in the UV. Moreover, no reliable data regarding the quantum efficiency of the setup over the required spectral region was available, thus making the present measurements necessary. It is worth noting that the procedure described here represents unique advancements made by the Quantum Optics Group at St Andrews toward using such apparatus to probe low intensity UV signals.

[7]Where the CCD spectrometer allows for measurements of the whole spectrum, the data acquisition system for the monochromator allows either the monitoring of the signal at a fixed wavelength in real time or scans to be taken over a wider range of wavelengths.

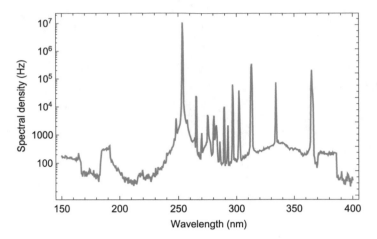

Fig. 5.15 Spectrum of the Tiramisu lamp from the OSC at St Andrews. Measured data

in their spectral and density properties. In what follows, we assess how well the monochromator can resolve these features. To do so, we set the width of the entrance and exit slits of the monochromators to equal aperture. The former is based on the Czerny-Turner design—see Fig. 5.16: a broadband illumination source (e.g., the light of the Tiramisu lamp, Fig. 5.12) is aimed at the entrance slit (A), which is placed at the focus of a curved mirror (C) that collimates the light and reflects it upon the grating (D). The collimated light is diffracted from the grating and collected by another curved mirror (E) which refocuses the light, now dispersed, on the exit slit (F). At the exit slit, the wavelength components of the broadband light are spread out—each wavelength arrives at a separate point in the exit-slit plane. The range of bandwidth transmitted through the exit slit is a function of the width of the slits.

We wish to determine the wavelength resolution of the monochromator. The spectral resolution, in nm, that is a function of the entrance and exit slits width (these are equally set), it the minimal resolvable wavelength difference. Because the width of the entrance and exit slits is always set equal, these will indifferently be referred to as the "slit width".

The measured spectra for increasing slit width inform us about the resolution of the monochromator over the wavelength range. The procedure to determine the spectral resolution of the monochromator is as follows: spectra acquired for large slit widths are fitted with a reference spectrum acquired for the narrowest slit width. The larger the slit width, the more light will enter the monochromator, thus the amplitude of the feature being scanned will increase. Furthermore, as was said earlier, the larger the slit width, the more details of all features become coarser and, ultimately, disappear. Therefore, the reference spectrum is fitted to the data by adjusting its amplitude and

Fig. 5.16 Photograph of the Czerny-Turner monochromator (Acton SpectraPro 2500i, Princeton Instruments Inc.). The main components of the monochromator are identified as **A** entrance slit, **B** plane mirror, **C** collimator (curved mirror), **D** grating (1800 g/mm grating), **E** focuser (curved mirror), and **F** exit slit

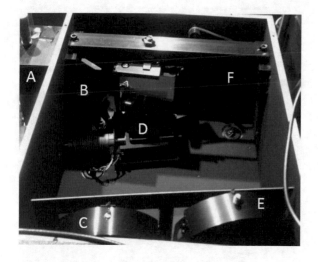

resolution—this is achieved by smoothing the reference spectrum with a Gaussian distribution of amplitude h and standard deviation σ

$$g(s) = h\frac{e^{-s^2/\sigma^2}}{\sqrt{2\pi}}. \tag{5.6}$$

The reference spectrum was taken with a slit width of 0.01 nm, smaller than the specified instrument resolution. The parameter σ for each slits width is the resolution of the monochromator for this slit width and wavelength (or wavelength range). I automated this fitting procedure in Mathematica, thus allowing for fast and reliable fitting and σ calculation. I varied the slit width between 0.01 and 0.5 mm, and could fit the 250 and 365 nm peaks with a Gaussian distribution (5.6) up to a slit width of 0.1 mm. For apertures wider than 0.1 mm, the spectrum was clearly distorted and spread with no recognizable features of the peak left. Likewise, the smoothing procedure failed to yield a reasonable resolution for the ramp measured with a 0.25 mm slit width—although the spectral features of it were still recognisable. Examples of the fit for each spectral feature is plotted in Fig. 5.17, for slits width of 0.05 mm. I also measured the spectral resolution of the monochromator at 532 nm by using light from the probe laser sent through the fibre (without an IR pulse).

The resolution decreases with slit width and levels for very small widths. The spectral resolution for the three features of the Tiramisu lamp and the 532 nm line are shown in Fig. 5.18. There is a difference between the resolution measured for the 192 nm ramp and the 250 and 365 nm peaks (those two are similar), as well as with the resolution at 532 nm. The latter difference may be because the grating is not designed to work identically at such dissimilar wavelengths. Princeton Instruments specifies a spectral resolution of 0.05 nm for the 1800 g/mm grating at 435.8 nm

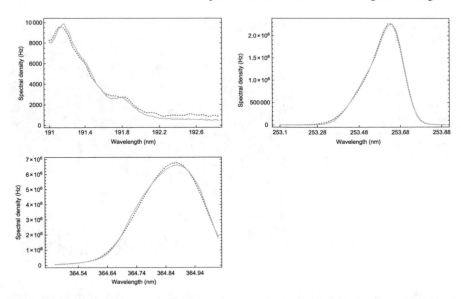

Fig. 5.17 Gaussian fitting of spectral features of the Tiramisu lamp. Raw data for a slit width of 0.05 mm is shown in dotted blue, and the smoothed high resolution spectra are shown in solid red for the 192 nm ramp and the 250 and 365 nm peaks

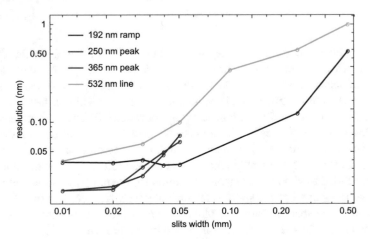

Fig. 5.18 Resolving power σ^{-1} of the monochromator (in nm^{-1}) for varying (equal) entrance and exit slit width (in mm). No data is shown for slit widths larger than 0.05 mm for the 250 and 365 nm peaks (in blue and red, respectively) because the feature shape and counts saturated for such large slits

and a 0.01 mm slit width—note that the value obtained at 532, 0.04 nm, is close to these specifications. The spectral resolution is clearly sub-nanometre for most slits opening.

5.2.6.2 Quantum Efficiency in the UV

In the previous paragraph, we established that narrow-linewidth UV signals can be resolved with a sub-nm resolution by the monochromator for narrow slit widths. In order to fully characterise the spectral sensitivity of the detection apparatus, we now need to estimate the quantum efficiency of the apparatus in detecting photons in this spectral feature. Any finite exit slit width s corresponds to a bandwidth $\Delta\lambda_s$. If the signal of wavelength λ to be detected has a bandwidth $\Delta\lambda \leq \Delta\lambda_s$, it is possible to formulate an estimation of the quantum efficiency from the reflectivity of the optics of the monochromator and detection efficiency of the SPC at the signal wavelength, as specified by the manufacturer. This can be easily confirmed experimentally by measuring the power of a narrow-linewidth signal on a power meter, and comparing it with the signal measured by the monochromator and SPC. If, however, the signal has a bandwidth such that $\Delta\lambda > \Delta\lambda_s$, there will be losses at the exit slit.

When carrying out the experimental assessment of the quantum efficiency of the detection apparatus, we did not have a quasi-monochromatic, narrow-linewidth, source that emits in the UV range over which we expect to observe the signal. Actually, the only source that emits at short-enough wavelengths was the DHS lamp. Light from the lamp transmitted through the UV filter has the appropriate wavelength, but is very broadband (218–238 nm FWHM, see Fig. 5.13). Only a fraction of the (broadband) light incident on the exit slit will enter the SPC. For a broadband source, these losses make the quantum efficiency artificially smaller. Thus, it is necessary to measure the losses at the exit slit to estimate the quantum efficiency of the detection apparatus.

Measuring a narrow-bandwidth feature of the spectrum for increasing slit width reveals an apparent broadening in the bandwidth of this feature. The broadening, in nm, is related to the change in slit width s by the dispersion of the grating and a convolution with a top-hat distribution—the transmission of the monochromator for this wavelength. In the previous section, we determined the spectral resolution of the monochromator, i.e., the bandwidth of the exit slit, for very narrow (equal) entrance and exit slit widths. Unfortunately, the signal is expected to have a very low intensity, and wider slit widths will have to be used to detect it. For example, in Sect. 5.4.1.2 we will present spectra of a peak at 260 nm that is expected to be much more intense than the signal: this peak was sufficiently strong to be observed for (equal) entrance and exit slit widths of 0.25 mm or larger, for which our earlier investigation in Sect. 5.2.6.1 did not yield conclusive spectral resolution. Thus, we now proceed to measuring the smearing of features of the (filtered) DHS lamp spectrum for very large slit widths, from 0.25 to 3 mm in order to infer the spectral resolution.

To that aim, one may calculate the transmission bandwidth of the exit slit, $\Delta\lambda_s = \alpha s$, that depends on the slit width s and the dimensionless scaling parameter α (of order 10^{-6}), that relates the slit width to the bandwidth of the slit. Smoothed or smeared spectra for large slit width s[8] are given by

---

[8]In (5.7), the spectrum is smoothed with a top-hat distribution of width $\Delta\lambda_s$, the transmission function of the slit.

$$R(\lambda) = \frac{1}{\alpha s_0} \int_{-\frac{\alpha s}{2}}^{\frac{\alpha s}{2}} R_0(\lambda') d\lambda', \tag{5.7}$$

where $R_0(\lambda)$ is a reference spectrum acquired for a narrow slit width $s_0 = 0.25$ mm. We remark that

$$R_0(\lambda) = S(\lambda) \Delta \lambda_{s_0} = S(\lambda) \alpha s_0, \tag{5.8}$$

with $S(\lambda)$ being the instrument limited spectral density.[9] From (5.7) and (5.8) we obtain α as

$$R(\lambda) = \frac{s}{s_0} < R_0(\lambda) >_{\Delta \lambda_s}, \tag{5.9}$$

where $< R_0(\lambda) >_{\Delta \lambda_s}$ is the average of the spectrum $R_0(\lambda)$ taken over the bandwidth $\alpha s = \Delta \lambda_s$. The transmission of the slit is then found by the ratio of the energy in the spectrum over $\Delta \lambda_s$, i.e., the integral over $\Delta \lambda_s$ of the large-slit spectrum (5.7), with the energy of the full spectrum

$$T_{\text{exit slit}}(\lambda) = \frac{\int_{-\frac{\alpha s}{2}}^{\frac{\alpha s}{2}} R_0(\lambda) d\lambda}{\int_{-\infty}^{+\infty} R_0(\lambda) d\lambda}. \tag{5.10}$$

As in Sect. 5.2.6, the above procedure was automated in Mathematica. We find an optimal fitting parameter $\alpha \approx 1.8 \times 10^{-6}$ for slit widths $s \geq 0.25$mm. Thus a slit of width $s = 3$mm (widest opening of the entrance and exit slits) covers a bandwidth of 5.25 nm. This yields $T_{\text{exit slit}}(229\text{nm}) = 11.1$ %: this is the fraction of the light from the broadband source that arrived at the exit slit that will reach the sensitive surface of the SPC. Finally, the *reduced* quantum efficiency of the setup with losses at the slit is calculated by multiplying the quantum efficiency of the SPC, $\eta_{SPC}(\lambda)$, with the transmission of the slit and the reflection efficiency of the monochromator:

$$\eta_{\text{reduced}}(\lambda) = \eta_{SPC}(\lambda) \, T_{\text{exit slit}}(\lambda) \, \eta_{\text{monochromator}}(\lambda). \tag{5.11}$$

Let us estimate the reduced quantum efficiency. The three mirrors of the monochromator are specified to have a reflectance of $R_{\text{mirror}} = 92\%$ around 229 nm, whilst the grating is specified to have $R_{\text{grating}} = 66\%$ reflectance at 225 nm. Thus, under the assumption that the incoming light is focused at the entrance slit and does not clip, the efficiency of the monochromator at 229 nm is

$$\eta_{\text{monochromator}}(\lambda) = R_{\text{mirror}}(\lambda)^3 \, R_{\text{grating}}(\lambda) = \left(\frac{92}{100}\right)^3 \times \frac{66}{100} = 50.6\,\%. \tag{5.12}$$

So only 50.6% of the (filtered) DHS light that enters the monochromator will reach the exit slit. When a measurement at 229 nm is performed, the exit slit will transmit

---

[9]The spectral density is limited in amplitude and bandwidth by the width of the slit, which sets the spectral resolution of the monochromator. See Sect. 5.2.6.

11.1% of this light through to the SPC: in total, only 5.62% of the input light will reach the SPC. Given the quantum efficiency of the SPC, 15% (see Fig. 5.14), this gives a reduced quantum efficiency, from entrance aperture to detection, of 0.84% at 229 nm when the entrance and exit slits of the monochromator are fully open to $s = 3$ mm.

Experimentally, we measure the amount of light from the DHS lamp transmitted through the UV-filter with a power meter, and get 52.3nW $\pm$ 5nW. With the monochromator ($s = 3$ mm) and SPC, we observe 5.24×10^8 counts per second at 229 nm, that is a power of 0.448 nW—yielding a quantum efficiency of 0.86%. This value is, within the error of the measurements[10] equal to that estimated in the previous paragraph. This study demonstrates that the quantum efficiency of detection for a broadband spectrum will be reduced by the monochromator spectral resolution (which is a function of the width of the entrance and exit slits). It also confirms that the reflection efficiency of the monochromator and the quantum efficiency of the SPC are close to those specified by the manufacturer—50.6 and 15 %, respectively.

Fortunately, we do expect the UV signal, of wavelength ≈ 220 nm, to have a very narrow linewidth. In this case, there will be virtually no loss at the exit slit, even for wide slit widths. In this case, the signal will be detected with a quantum efficiency of $\eta_{\text{quantum}}(\lambda) = R_{\text{monochromator}}(\lambda) \times \eta_{SPC}(\lambda) \approx 7.6$ % for $\lambda = 220$ nm. Given this low quantum efficiency of the detection apparatus, it is clear that we cannot afford to lose light in the UV by other filtering techniques than the CRF introduced in Sect. 5.2.5.1. In particular, resorting to the 228 nm bandpass filter is significantly worse, for it transmits a maximum of 25% of the incident light across its bandwidth.

To conclude, in this section of the dissertation, we have presented the investigation of the spectral resolution of the detection apparatus, and found that we could measure spectral features of incident light with sub-nm precision for very narrow slit widths, and a quantum efficiency of 7.6% (for wide slit widths) in the deep UV region, where parametric amplification in the negative-norm mode occurs. This brings our journey on the optical table to an end: we now know all the details of the experimental setup assembled for this thesis and can proceed to looking into the results of the experimental observation of the scattering of a positive-norm mode on the sides of a fundamental soliton in the fibre.

5.3   Scattering to a Positive-Norm Mode

In this section of the dissertation, we discuss the phenomenon of scattering from a *in* mode α_0 of positive norm to and *out* mode α_1 of positive norm, and present the original experimental results that demonstrate this effect. The main two experiments relevant to the present considerations were carried out in St Andrews and presented

---

[10]Note that these measurements fluctuate in time, thus their precision is not better than 5%, and the counts are collected over a wide, 5.25 nm, bandwidth.

in [1, 2], for the scattering of an IR and a visible probe, respectively, on a soliton generated by an IR pulse.

5.3.1 Positive-Norm Scattering Efficiency

In this Thesis, a visible, CW, probe is scattered on a soliton in the PCF. The efficiency of this scattering process is ruled by Eq. (5.4), where the amount of of light in the *in* coherent mode depends on two parameters: the magnitude of the refractive index change under the soliton and the velocity difference between the *in* mode and the soliton. As for the former, as we already discussed in Sect. 2.2.2 of this dissertation, in the scattering process, the comoving frequency ω' is a conserved quantity, and so the probe frequency ω follows a contour line of ω' as a function of the nonlinear index induced by the pulse—see Fig. 2.5. The nonlinear susceptibility experienced by the soliton pulse at the carrier wavelength λ_0 is [1, 12]

$$\delta n = \frac{c \lambda_0 D_0}{(\omega_0 T_0)^2} \tag{5.13}$$

where D_0 denotes the dispersion parameter at λ_0, and ω_0 is the carrier frequency. As was shown in Sect. 4.3, any increase δn in the refractive index of a dispersive medium leads to mode mixing at the interfaces between regions of low and high refractive index—and thus to parametric amplification, transfer of energy to the *out* mode.[11] However, all of the light in the probe wave might not be able to scatter at the interface. Indeed, because the group velocities of the probe $v_g(\omega_{probe})$ and of the pulse u are similar, only a small fraction of the total probe light can be converted within the finite length of the fibre. Thus, the amount of energy available for the scattering process, that is, the fraction of the probe power that can be scattered into outgoing modes (of positive and negative norm alike) is dependent on the repetition rate, the length of the fibre and the inverse of the difference between u and $v_g(\omega_{probe})$ [1]:

$$\eta_{int} = \nu_{rep} L \left(\frac{1}{u} - \frac{1}{v_g(\omega_{probe})} \right). \tag{5.14}$$

For our experimental setup, η_{int} is maximally on the order of 10^{-4} [2].

The effect of frequency shifting is also described by the theory presented in this thesis. Consider a RIF, as schematised in Fig. 3.4, that models the leading edge of a soliton in the fibre. As we saw in Sect. 3.2.3, there exists a frequency interval over which a mode in which light propagates towards the RIF scatters into a mode in which light propagates away from the RIF (the uniquely escaping mode *moR*).

---

[11]Note that in this dissertation, we sometimes adopt the expression "frequency shift" to describe parametric amplification of an *out* mode of positive-norm. This is inherited from the language of the community (see for example [1]).

A symmetrical configuration exists if the RIF models the falling edge of the soliton. However, when considering the process of scattering to a single positive-norm mode, it is actually possible to use a simple, dispersion-less, tunnelling model that implements the NLSE (2.62) to determine the efficiency of the probe-pulse interaction [2]. The model [2] has the advantage of readily allowing for the study of smooth pulse profiles. A derivation of the quantum tunnelling of a wave at a smooth interface in a dispersion-less medium is provided in Appendix C of this dissertation, for the sake of conciseness only the main results and phenomenological arguments will be presented here.

5.3.2 Tunnelling Model for Probe-Pulse Interaction

Here I present the analytical theory of scattering of light at solitons in fibres, including frequency shifts and wave tunnelling. In the frame moving with the soliton, that is assumed to be unaffected by the probe and has an amplitude of the form $P_0 \, \text{sech}^2(\tau/T_0)$, the NLSE (2.62) can be written as [2]

$$\frac{\partial^2 A_1}{\partial \tau^2} - \frac{2}{\beta_2} \left(\beta_1 (\omega' - \omega'_m) + r\gamma P_0 \, \text{sech}^2(\tau/T_0) \right) A_1 = 0, \qquad (5.15)$$

where A_1 is the amplitude of the probe wave, τ is the retarded time (see 2.62), β_1 and β_2 are the first and second derivatives of the propagation constant $\beta(\omega)$,[12] γ is the fibre nonlinearity and r is a factor accounting for the reduction in cross-phase modulation due to conditions such as the relative polarization orientation or mode size mismatch. The analogy of (5.15) with the Schrödinger equation in quantum mechanics allows to investigate quantum mechanical problems with classical fibre optics.

To the probe wave, the soliton is a constant one-dimensional potential, for which the transmission and reflection coefficients, T and R, can be found. For a step-like potential these are the β coefficients of Sect. 3.2.4. For a hyperbolic secant squared potential they are (see Appendix C):

$$T = \frac{1}{1+\xi} \quad R = \frac{\xi}{1+\xi},$$

$$\xi = \begin{cases} \dfrac{\cos^2(\pi/2\sqrt{1-B})}{\sinh^2(\pi(\omega-\omega_m)T_0)} : & B < 1 \\[2ex] \dfrac{\cosh^2(\pi/2\sqrt{B-1})}{\sinh^2(\pi(\omega-\omega_m)T_0)} : & B \geq 1. \end{cases} \qquad (5.16)$$

The transmission through the soliton—the potential barrier—is therefore determined by only two parameters: the ratio of detuning to the soliton bandwidth $(\omega - \omega_m)T_0$,

---

[12] β_1 is the inverse of the group-velocity of the probe and β_2 is the GVD parameter at the probe wavelength, see Sect. 5.2.3.

and the normalized barrier height $B = 8r\gamma P_0 T_0^2/\beta_2$. In [2], F. König found that effective reflection on the barrier can be achieved for very large detuning, and is not limited by the spectral width of the soliton-pulse (provided that B is sufficiently large). Contrarily to the photon picture of Four Wave Mixing (FWM), the mode conversion at stake here is a collective effect of the modes of the soliton and the probe, and not a phase matched mixing of only four modes. Indeed, the barrier height required for frequency shifting was found to increase quadratically, and not exponentially as FWM would require.[13] This is a remarkable result: frequency shifting at the soliton-edge is a feature of horizon physics for which no simple alternative explanation (such as FWM) can be provided by nonlinear optics. Considering the efficiency of reflection of a CW probe at the soliton, the conversion efficiency R (Eq. 5.16) is reduced by η_{int}, the fraction of the probe light that interacts with the soliton, to the total efficiency η_{tot}. Phenomenologically, one sees that: the larger the detuning of the probe from the group velocity of the soliton, the more light collides with the soliton with higher relative speed. For small detunings there is negligible tunnelling and the probe is nearly perfectly reflected. For larger detunings, all the probe light tunnels through the soliton and B, the height of the barrier required for efficient frequency shifting, decreases quadratically.

5.3.3   Visible Frequency Shifts at the Horizon

The findings of the dispersion-less tunnelling model shed light on the physics of frequency shifting at the horizon. In this process, an *in* mode of positive norm scatters at the refractive index front (RIF), into an *out* mode of positive norm and positive group velocity in the moving frame. Because of the Doppler shift, this *out* mode does not have the same laboratory frame frequency ω as the *in* mode, although they share the same moving frame frequency ω'—energy is conserved in the moving frame. Since the *in* and *out* modes have a different laboratory frame frequency, it is commonly said that the *in* mode was frequency shifted by the RIF—and, indeed, if all of its light were made to scatter on the RIF, for appropriate RIF height, light coming out of the fibre would only be in the *out* mode. And, actually, in the experiment, the output light is measured after remaining light of the *in*-mode-frequency has been filtered. From the theory presented in the previous section, it is possible to calculate both the efficiency of this frequency shifting effect and the wavelength at which light in the *out* mode will be observable in the laboratory.

Because of the nature of the setup used in the experiments presented here, a large fraction of the frequency shifted light could not be measured. Thus, measurement of the efficiency of the scattering of positive-norm to positive-norm light at the soliton was not made. However, the experimental results presented here allow for extending the mapping of the frequency shifting as a function of detuning. This frequency

---

[13]In particular, the edges of the curve of efficiency of the frequency shifting effect as a function of the soliton-probe detuning are different from the exponential fall one would obtain for FWM [2].

detuning is achieved by tuning the central wavelength of the pulse (see Fig. 5.3), i.e. the soliton velocity. The wavelength that the probe will shift to is determined by the fibre dispersion and the conservation of ω', by

$$\delta\lambda = L\nu_{rep}\beta_2(\omega - \omega_m). \tag{5.17}$$

$\delta\lambda$ depends on the relative velocity of the probe and soliton, according to dispersion, as well as on the interaction length. The dispersion relation for the PCF is based on an approximate calculation of the group index obtained from a silica strand model [2] which reproduced the group velocity matching condition in [10]. The dispersion relation was then fitted to a Sellmeier model for dispersion. We found that the best fit was obtained for a material with two resonances in the IR. We checked that the Sellmeier equation (3.59) thus enforced allowed for reproducing the theoretical predictions of [8, 10]—and indeed, we find a central wavelength for the NRR of 224 nm, where J. McLenaghan observed it at 222 ± 1 nm.[14] Furthermore, the prediction of the wavelength of light shifted from 532 nm matches that of [2]: we obtain a shifted wavelength that depends approximately hyperbolically on the central wavelength of the soliton, as shown by the (calculated) red curve on Fig. 5.19.

In the experiment, the group velocity of the soliton is set by its centre wavelength in the dispersive fibre. I tuned this wavelength to realise situations where the probe wave (at $\lambda = 532$ nm) is slower than the soliton and is overtaken by it, and vice versa. Figure 5.20 displays two epitome spectra of the frequency shifted probe light. These spectra correspond to a -9 and $+13$ nm spectral shift of the probe. The spectral

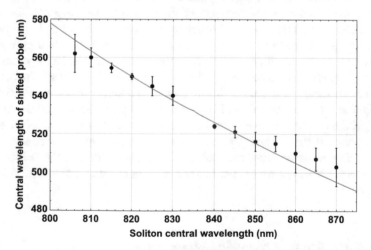

Fig. 5.19 Location of shifted probe spectra for different soliton wavelengths. The red curve is the prediction from the dispersion curve of the PCF. The wavelength of the shifted probe spectra was measured with a precision that depends on reflection on the various optics, and the complicated features that depend on Raman interaction and higher order dispersion [13]

---

[14]See Sect. 2.2.2 for a presentation of the underlying theory.

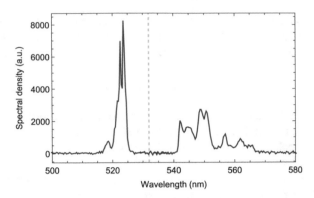

Fig. 5.20 Two spectra of the blue and the red shifted probe light. The input light initially was centred at $\lambda = 532$ nm, but filtered in the output by means of a notch filter (see Fig. 5.11 for its transmittance). The soliton was tuned to 845 and 825 nm, respectively. Spectral shifts of -9 nm (blue) and $+13$ nm (red) are observed. The relative spectral density are not representative of the efficiency of the frequency shifting process for they are distorted by clipping of the beam on the optics and the filtering effect of the notch filter

width and structure depend on parameters such as the detailed pulse shape—which is affected by Raman interaction and higher order dispersion [13]—and remains to be fully explained.

I repeated this experiment with various pulse wavelengths, from 749 to 887 nm (see Fig. 5.3) to map out the frequency shifting as a function of detuning. Figure 5.19 shows the measured centre wavelength of the shifted probe wave as a function of soliton wavelength. These results further those presented in [2] and are the most extensive map to date. Note that the centre wavelength of the shifted light follows the condition set by the dispersion of the fibre, thus the soliton had approximately constant group velocity, unaffected by higher order dispersion. The spectra presented here are remarkable: as stated previously (see the argument following Eq. 5.16) the input light at 532 nm was red- and blue-shifted to up to 560 nm and down to 505 nm, respectively—that is, over a maximum of 28 and 27 nm, which is 1.8 times the bandwidth of the soliton! This comes as an experimental confirmation of the understanding we drew from the theory: frequency shifting at the soliton is not a mere manifestation of four wave mixing but a genuine and signature feature of horizon physics.

5.4   Scattering to a Negative-Norm Mode

It is possible to carry out a similar analysis to that presented in the previous section to calculate the wavelength at which light in the negative norm mode will be observed: for a given carrier frequency, the group velocity of the pulse that generates the soliton in the fibre will depend upon the dispersion of the fibre. After a Lorentz boost, the

Fig. 5.21 Diagrammatic phenomenology of scattering at the soliton edge (black *sech*^{2} profile) in the comoving frame. Energy in the positive-norm *in* mode *moL* (*uoR*) scatters at the soliton into an *out* mode of positive norm, *uoL* (*moR*), and an out mode of negative norm *noL*

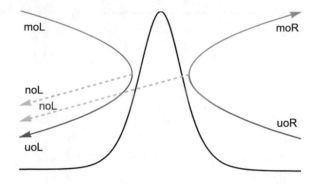

frequency shifts to $\omega' = \gamma(\omega_{\text{probe}} - uk)$ in the moving frame, with u the velocity of the frame (taken to be $u = v_{\text{gpulse}}$). Energy transfer from *in* to *out* modes occurs for constant comoving frequency ω'. The conservation of energy in the moving frame translates to a contour line of slope u and ordinate at origin $\omega = \omega'$ (as exemplified on Fig. 3.3 by the blue contour line). Considering scattering at the front (back) of the soliton, light in the positive-norm, coherent, *in* mode *uoR* (*moL*) will be red-(blue-) shifted to the positive-norm *out* mode *moR* (*uoL*) that is an oscillatory mode in the same refractive index region as the *in* mode—that is on the low refractive index side of the interface—as diagrammatically depicted on Fig. 5.21. In addition, Eq. (5.4) states that some energy will be transferred to an *out* mode of negative norm, *noL*, by parametric amplification. This negative-norm *out* mode allows for light to propagate away from the soliton in the low-refractive-index region on the left of the soliton, as in Fig. 5.21.

Scattering into *noL* is allowed because the contour line of constant ω' that passes through the point of frequency ω_{probe} (the frequency of the 532 nm cw probe wave in the laboratory frame) intersects with the dispersion branch of negative-laboratory-frame optical frequency, as is exemplified in Fig. 3.3. More precisely, Fig. 5.22c shows the negative-laboratory-frequency optical branch, and two such ω' contour lines, for the extremal central pulse wavelengths attained in the experiment, 800 nm in (a) and 865 nm in (b). A theoretical prediction, the purple curve on Fig. 5.22d, shows the calculated wavelength of mode *no* over the whole range. The wavelength at which light in the negative-norm mode will be observed, in the neighbourhood of 220 nm, varies by less than 7 nm for the full range of soliton wavelengths. Note that this wavelength range is independent of the change in refractive index under the soliton, for the *out* mode of negative-norm allows for light to propagate away from the soliton in the low refractive index region. This is in contrast with the study of the soliton edge, modelled as a step, presented in Chap. 4, according to which the *out* mode of negative norm allows for light to propagate away from the refractive index front (RIF) in the high refractive index region. Thus, although it is possible to calculate the expected output wavelength (see Fig. 5.22d), the efficiency of the parametric amplification process described by Eq. (5.4) cannot be calculated by using a step in

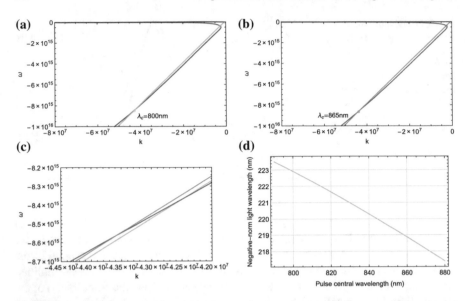

Fig. 5.22 a–d: Determination of the wavelength of light in the optical negative-norm mode *noL* as a function of the wavelength of the soliton. The branch of negative optical-laboratory-frequency ω is shown in blue. Contour lines of constant comoving frequency $\omega' = \omega'_{\text{probe}}$ are shown for the extremal central wavelengths of the soliton generated in the experiment: **a** $\lambda_c = 800$ nm, and **b** $\lambda_c = 865$ nm. An excerpt of the branch of negative optical-laboratory-frequency is shown in (**c**) to illustrate the bandwidth across which the $\omega' = cst$ line sweeps along the branch. The wavelength at which light in *noL* should be observed is plotted against the central wavelength of the pulse in (**d**)

the refractive index—a more complex and realistic profile is needed and this was not at all considered for this Thesis.

In the moving frame of the soliton, a nonliinear refractive index region is surrounded by two asymptotic regions of linear index. This setting is typical for pulses propagating in dielectric media. Such a configuration is illustrated in Fig. 5.21: mode *uoR* transfers energy to mode *noL* that allows for light to propagate on the *other side* of the soliton. Thus, one would intuitively expect the efficiency of the scattering to depend upon the change in refractive index induced by the soliton, as well as on the soliton pulse-length. A more quantitative prediction is difficult to make without careful study of the dispersion relation and thorough implementation of the algorithm presented in Chap. 4 for a parsed profile. For example, one would expect that reflection upon the interior edges of the soliton would yield an etaloning effect in the spectrum of modes transmitted through the high refractive index region under the soliton to the low refractive index region on the opposite side of the soliton.[15]

---

[15]Likewise, it would not be surprising if the scattering efficiency depended upon the edge of the soliton profile with which the probe would interact: energy transfer might be more efficient if the probe interacts with the back than with the front of the soliton, for scattering into *no* would then resemble a reflection process (whereas it would resemble a transmission process, through the 'etalon', if the probe would scatter on the leading edge of the soliton).

Thus, in what follows, a scattering coefficient $\left|\beta^{\text{probe},no}\right|^2 \approx 10^{-10}$, a preliminary calculation with a soliton is used as a quantitative guide in the search for the negative-norm signal.

5.4.1 Signal to Noise Ratio in the UV

As in all experiments, the measurements we perform are limited by the signal to noise ratio (SNR), where the signal is the detectable rate and the noise presently is the dark counts and the background (light from the laboratory and from the green and IR components of the beam). In this section, we use the techniques presented in Sect. 5.2.5 to improve the SNR by filtering the noise strengther than the signal.

5.4.1.1 UV Signal Strength

Let us first estimate the signal strength. As in Sect. 5.3, we assume a pulse-probe interaction efficiency $\eta_{int} = 10^{-4}$. This is combined with Eq. (5.4) to calculate the power scattered from the *in* mode to the *out* mode:

$$P_{UV} = \eta_{int}\left|\beta^{\text{probe},no}P_{\text{probe}}\right|^2 \approx 10^{-4} \times 10^{-10} \times 60 \times 10^{-3}\text{W} = 0.6\text{fW}, \quad (5.18)$$

where we assumed that 60 mW of green light propagate in the fibre. This power is limited by the coupling efficiency of 532 nm light in the fibre, and ultimately by the amount of input power available (≤ 300 mW with the current setup). Assuming negligible fibre losses in the UV, the expected photon rate at the output of the fibre is

$$\eta_{no} = \frac{P_{no}\lambda_{no}}{hc} \approx \frac{60 \times 10^{-17}\text{J.s}^{-1} \times 220 \times 10^{-9}\text{m}}{6.626 \times 10^{-34}\text{J.s} \times 3 \times 10^8\text{m.s}^{-1}} = 664 \text{ Hz}, \quad (5.19)$$

with h the Planck constant and c the speed of light in vacuum.

The total quantum efficiency of the setup over the emission range of *no* is set by the product of: the transmission efficiency of the UV-condenser triplet, the transmission efficiency of CRF, the reflection efficiency of the corner-UV mirror, the transmission efficiency of the focusing lens and the quantum efficiency of the detection apparatus, see Sect. 5.2.5. These quantities are summarised in Table 5.1. Recalling that the SPC has a quantum efficiency of $\approx 15\%$ at 229 nm, we obtain a total detection efficiency $\eta_{\text{detection}} \approx 3.4\%$ for our setup. According to the output power calculated in Eq. (5.18), this means that the signal strength, i.e. the registered count rate of the signal, is

$$S_{no} = \eta_{no}\,\eta_{\text{detection}} = 20 \text{ Hz}. \quad (5.20)$$

Table 5.1 Transmission and reflection efficiency of the optical elements in the UV. Except where stated otherwise, the efficiencies are specified for $\lambda = 229$ nm at normal incidence

Optical element	T (%)	R (%)
UV-condenser triplet	89	–
UV mirror	–	92.3
UV mirror (45° incidence)	–	95
CRF (6 bounces)	62	–
UV lens	84	–
Monochromator mirrors	–	92
Monochromator grating (at 225 nm)	–	65
Monochromator (3 mm slits width)	50.6	–
UV-filter	20	–

With this signal strength, care has to be taken in the eperimental layout to avoid filters with a low signal transmission. The probe being essentially a single frequency mode, we expect light in *no* to be concentrated in a relatively narrow peak in the spectrum, which should allow for isolating this very weak signal from the noise.

5.4.1.2 Background Counts in the UV

We now seek the best SNR for the signal strength calculated in the previous paragraph. As was discussed in Sect. 5.2.5.1, we have tested a number of filtering techniques to reduce the background counts in the UV. In addition to direct filtering of the beam, the detection apparatus was physically isolated from the rest of the laboratory by means of a light-tight box that lets only the beam in. As a result of this physical "boxing", the laboratory background is $N_b < 1$Hz over the UV range, see Fig. 5.23. This background is measured by coupling the IR pulse and 532 nm probe into the fibre and blocking the end of the fibre so that no light escapes from it for slit width of 3 mm—thus only light scattered on the various surfaces of interaction with the input beams can possibly reach the detector. With this background alone, the SNR would be

$$\text{SNR}_{\text{background}} = \frac{S_{no}}{N_b} > 20. \tag{5.21}$$

Unfortunately, as was mentioned earlier, the remnant of the green light scatters off the grating and increases the background in the UV-measurement.

The 532 nm probe light is not fully filtered between the fibre and the entrance slit of the monochromator. In the monochromator, this uncollimated light scatters off the grating and creates counts at all wavelengths in the UV, as can be seen in Fig. 5.24. It shows the evolution of this "green background" as a function of the diameter of the iris set before CRF. The green background decreases as the iris is closed, as we would expect from spatial filtering. The green background is attenuated by about one order

Fig. 5.23 Minimal background counts in the UV. These counts are due to the glow of the laboratory itself—that is, to light that reaches the detector after reflection/refraction of the intense IR and green input beams on various surfaces in the laboratory and the dark counts

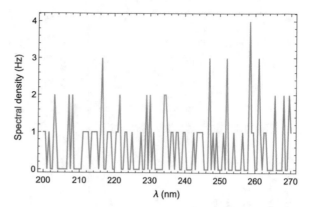

of magnitude from an iris diameter of 15 to 2.5 mm. Note that as the background decreases, a peak at 266 nm emerges—corresponding to second harmonic generation from the strong CW light at 532 nm. This is remarkable: this weak SHG, although not phase-matched, has a rate $S_{SHG} = 20$ Hz in all four data sets. This shows that the iris does only filter out the green light an lets the UV components unattenuated. So, due to this filtering process, the SNR has increased by one order of magnitude. The TH peak is at wavelengths much longer than the emission range of *no* and does not affect observations at shorter wavelengths. The green background never disappears at shorter wavelengths, and can be higher than the expected signal strength for large iris diameters—thus we shall use small iris diameters in order to improve the SNR.

In contrast to the green light, the IR light in the beam does not create any increase in the background counts, as can be seen in Fig. 5.25. According to the theory of Sect. 5.3.2, an increase in the soliton energy reduces the tunnelling probability and increases the coefficient of scattering to a negative norm mode, $\left|\beta^{\text{probe},no}\right|^2$. Although Sect. 5.4.1.2 was using fundamental solitons, we also use $N > 1$ solitons, with average powers of 2 mW ($N = 1.26$), 4 mW ($N = 1.79$) and 8 mW ($N = 2.53$) to increase the signal strength. As shown in Fig. 5.25, we observe a narrow peak at 262 ± 1 nm wavelength. This peak is the third harmonic (TH) generated from the soliton. For $\lambda_{TH} = 262 \pm 1$ nm, the fundamental lies at 796 ± 3 nm. We consistently observe TH at this wavelength, which implies that its generation relies on the bandwidth of the pulse (which is ≈ 10 nm, see Fig. 5.3). The TH is generated at wavelengths longer than the range of emission of *no* and does not affect the SNR for *no* detection: considering the IR background only, $N_{IR} = 1$Hz

$$\text{SNR}_{IR} = \frac{S_{no}}{N_{IR}} = 20. \tag{5.22}$$

To summarise, background from the laboratory is negligible (with an average rate below 1 Hz at all wavelengths). Likewise, the IR does not deteriorate the SNR in the UV. We note, however, the presence of a peak, due to THG from the IR pulse, at a wavelength far from the region of interest. Only the green probe creates counts in

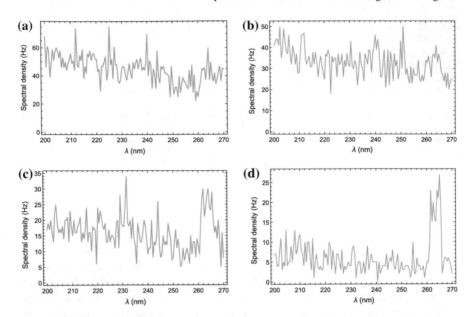

Fig. 5.24 Green background—Background in the UV due to the scattering of the green light off the grating in the monochromator. Spectra are shown for varying iris diameters: **a** 15 mm, **b** 10 mm, **c** 5 mm and **d** 2.5 mm for a 532 nm CW power of 50 mW in the fibre. The minimal background counts without incident light on the monochromator and SPC are shown in Fig. 5.23. Note the change of scale from (**a**) to (**d**): the green background decreases and lets a 20 Hz peak at 266 nm emerge

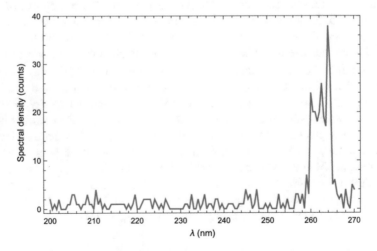

Fig. 5.25 IR background—background in the UV due to the propagation of IR light alone in the fibre. Average power of 8 mW for a central wavelength of 806 nm. Iris diameter of 15 mm. The background between 200 and 258 nm is indistinguishable from detector dark counts (cf. Fig. 5.23)

the UV, which can be limited to rates below 20 Hz by closing the iris down to 5 mm, without loosing genuine UV counts (such as the very well defined second harmonic generation peak). This yields a SNR of

$$\text{SNR}_{UV} = \frac{S_{no}}{N_b + N_{IR} + N_G - N_d - N_d} \approx 1 \qquad (5.23)$$

where we have subtracted the dark counts twice. Over the range 218–223 nm a signal of 20 Hz would be observed with a signal to noise ratio of 1. Closing the iris further, to 2.5 mm, yields $\text{SNR}_{UV} \approx 3.3$. The emergence of the SHG peak from the CW probe, as shown in Fig. 5.24, seems to indicate that the UV beam is not significantly attenuated by such a small iris—we may readily use this spatial filter to better the SNR.

5.4.2   UV Spectra

In this section, I will present the results of the experimental investigation of pulse-probe interactions. As was shown earlier, the propagation of the IR pulse in the fibre does not create a background in the UV would lower the SNR at the signal wavelength, even when the average IR power is such that a $N > 1$ soliton is generated in the fibre. Generating a high order soliton might be an interesting path to a higher signal to noise ratio (SNR). Indeed, as was said earlier, the increase in the refractive index under the soliton is larger than for $N = 1$ solitons, which effectively increases the height of the potential barrier (yielding a larger scattering coefficient). However, I also observe that the propagation of a higher order soliton in the fibre may lead to third harmonic generation (THG), resulting in the appearance of a narrow peak at large UV wavelengths (beyond 260 nm). It is necessary to assess the effect that the scattering of the probe on the IR pulse may have on THG, to rule out any contamination of the SNR at the signal wavelength.

5.4.2.1   Third Harmonic Generation

Let us first study the power dependence of third harmonic generation (THG): the evolution of the third harmonic (TH) peak as the average power of a pulse centred at $\lambda_c = 806$ nm is increased is shown in Fig. 5.26. For this central wavelength, the $N = 1$ soliton has a bandwidth $\Delta\lambda_c = 10$ nm. For N= 1, in (b), there is a very narrow spectral feature ($SNR = 6$) at $\lambda_{TH} = 266$ nm, that broadens and increases in amplitude as the IR power increases. Note that the peak shifts to longer wavelengths as the soliton order increases: e.g., for $P_{\text{avg}} = 4$ mW (e)), N ≈ 2, THG occurs at $\lambda_{TH} = 275$ nm. This clearly demonstrates the extreme dependence of THG on phase-matching and group-velocity matching.

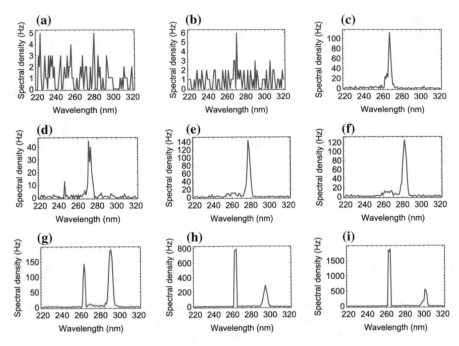

Fig. 5.26 Evolution of the third harmonic (TH) peak as the average power of a pulse centred at $\lambda_c = 806$ nm is increased. Spectra for power **a** 0.5 mW, **b** 1 mW, **c** 2 mW, **d** 3 mW, **e** 4 mW, **f** 5 mW, **g** 6 mW, **h** 7 mW, and **i** 8 mW are shown. Note the change in the scale of the spectra. The minimal background counts without incident light on the monochromator and SPC are shown in Fig. 5.23. The background created by the IR in the UV is of maximum 2Hz

Considering, Fig. 5.26c, the THG pulse generated at $\lambda_{TH} = 266$ nm has a bandwidth of $\Delta\lambda_{TH} = 5$ nm, that is actually the bandwidth of the slit open at 3 mm. Thus we cannot determine the pulse duration from this data. However, we can assume that the generated THG pulse width is comparable to the pump pulse [14], but, because of dispersion, the two move at different speeds through the fibre. Because of this group-velocity mismatch, the spectrum of third harmonic may not be located at exactly $\lambda_c/3$. To understand this, let us study the phase-matching condition for THG: for a quasi-CW pump launched at frequency ω_c, this takes the form [14]

$$\Delta\beta \equiv \beta_{TH}(3\omega_c) - 3\beta(\omega_c) = (3\omega_c/c)\,(\bar{n}_{TH}(3\omega_c) - \bar{n}(\omega_c)) = 0, \qquad (5.24)$$

where $\beta(\omega)$ and $\bar{n}(\omega)$ are the (frequency-dependent) propagation constant and effective mode index, respectively. The phase-matching of Eq. (5.24) implies that $\bar{n}_{TH}(3\omega_c)$ has to match $\bar{n}(\omega_c)$. This is only possible if the TH propagates in a higher-order transverse mode, which only occurs if the difference between the refractive index at ω_c and that at ω_{TH} is less than the core-cladding index difference—a quantity which is of the order of 0.1 for most PCFs [11]. If, now, we consider an ultrashort pump which is rather broadband, like the pulses in the experiment, we should expand the propagation constants of Eq. (5.24) for the pulse carrier frequency, $\beta(\omega_{IR})$, and the central frequency of the THG pulse, $\beta_{TH}(\omega_{TH})$, in Taylor series around those frequencies. Retaining terms up to first order in these expansions only, we find that [14]

$$\Delta\beta = \beta_{TH}(3\omega_c) - 3\beta(\omega_c) + 3\,(\omega - \omega_c)\,\Delta\beta_1, \qquad (5.25)$$

where $3\,(\omega - \omega_c)$ is the frequency shift of third harmonic and $\Delta\beta_1 = v_{gTH}^{-1} - v_{gIR}^{-1}$ is the GDD between the pump and the TH frequencies. The condition $\Delta\beta = 0$ is satisfied when TH is shifted from $3\omega_c$ by $-(\beta_{TH}(3\omega_c) - 3\beta(\omega_c))/\Delta\beta_1$. This shift depends on the group-velocity mismatch. In Fig. 5.26 we see that TH is shifted to longer wavelengths (shorter frequencies). We also observe that the THG spectrum exhibits two distinct peaks for high average IR power. This is because the $N > 1$ soliton fissions into two pulses (as in Fig. 5.10), one of which moves to longer wavelengths than the fundamental, to which the longer-UV TH is phase-matched.

Equation (5.25) shows how broad the spectrum of the fundamental has to be to contain phase-matched fundamentals. However, although both the propagation constants of the pulse and the TH were developed in Taylor series to arrive at this result, thus keeping $\omega_{TH} = 3\omega_c$, it can also be interpreted as follows: for low pulse energies, $(\beta_{TH}(3\omega_c) - 3\beta(\omega_c)) = 0$ and the process is phase-matched. Increasing the power, however, creates different nonlinear contributions in β_{TH} and β. Thus $(\beta_{TH}(3\omega_c) - 3\beta(\omega_c))$ is no longer 0 and (5.25) shows that a slightly different frequency becomes phase-matched, namely $\omega = \omega_c - (\beta_{TH}(3\omega_c) - 3\beta(\omega_c))/3\Delta\beta_1$. This might account for slight discrepancies between the theoretical phase-matched wavelengths of either of the TH peaks and the measured wavelengths.

Note that these spectra exhibit no feature at short UV wavelengths, where the noise (due to the laboratory background and IR background, see Sect. 5.4.1.2) is of maximum 2 Hz, as can be seen on Fig. 5.26a. Thus, although it may shift in wavelength as a function of the fibre mode in which it propagates, TH will not affect the SNR at the signal wavelength. I also verified that TH exponentially disappears for longer pulse wavelengths: for $\lambda_c > 820$ nm, no TH peak was observed, even for high average IR powers of the order of 8 mW. Beyond these observations, I established that THG also depends on the coupling efficiency of the input IR beam in the fibre. Indeed, for the pulse central-wavelength regime at which TH was normally observed (790–820 nm), if the coupling efficiency was decreased to only 10% (by misaligning the input beam on the tip of the fibre), no TH peak was observed. This is because the IR mode then mostly overlaps with the cladding and not with the core, and the phase-matching condition (5.25) cannot be fulfilled because $\bar{n}(3\omega_c) - \bar{n}(\omega_c) < 0.1$. This comes as a complement to the argument drawn in Sect. 5.2.3.2, in which the average IR power necessary to generate a fundamental soliton was found to be higher than in the calculations that assume that all the IR power was confined to the core. Thus, lowering the coupling efficiency of the IR beam in the fibre might be another means to better the SNR at the signal wavelength if we find that the interaction of the probe with higher-order solitons lowers the SNR.

5.4.2.2   Pulse-Probe Interactions

It is finally time to scan the UV when the 532nm CW light is made to interact with a fundamental soliton in the IR. In the experiment, I scanned a four-dimensional parameter range: the central wavelength and average power of the IR pulse could be varied independently, the probe power could be varied, and the diameter of the iris in front of CRF could be varied. In Fig. 5.27, a UV spectrum for a $P_{avg} = 8$mW pulse at $\lambda_c = 865$ nm interacting with a 33 mW, 532 nm, CW probe is shown. Above the background created by the green beam ($b_g = 30$HZ) we clearly see a $S_{247} = 15$ Hz peak of bandwidth $\Delta\lambda_{247} = 5$ nm at $\lambda = 247$ nm. The signal to noise ratio (SNR) in this configuration is only $SNR_{865} = 2$, but the peak is clearly visible. The SNR could be improved by subtracting the green background from the spectrum. However, we note that the level of this green background is lower than that recorded for an iris of 15 mm diameter. Moreover, the SHG peak from the green probe is not observed in this spectrum. This is because the probe power is lower for this measurement—down from 50 mW in Fig. 5.24 to 33 mW. Thus the energy available for SHG is lower and the peak remains hidden in the 15 Hz background.

No signal at the expected wavelength ($\lambda_s \approx 220$ nm) can be seen in Fig. 5.27. Furthermore, the peak at 247 nm ("mid-UV-wavelength peak") is intriguing: no obvious phase-matched process yields a peak at this wavelength. Note that its spectral shape is likely given by the transmission of the monochromator slits: it is characteristic of a top-hat function. Recalling the signal strength calculated in Sect. 5.4.1.1, S_{no}, we note that it is similar to the amplitude of this mid-UV-wavelength peak. For $\lambda_c = 865$ nm, the 532 nm probe is slower than the pulse and thus interacts with its leading edge. Unfortunately, the interaction of the probe with the leading edge of other long-wavelength pulses did not yield a similar peak across the parameter range, i.e., I did not observe the mid-UV-wavelength peak for 835nm $< \lambda_c <$ 887nm, except at $\lambda_c = 865$ nm, for $P_{avg} = 8$ mW and an iris diameter of 15 mm.[16]

On the other hand, the peak was consistently observed for short-wavelength pulses, such that the probe interacts with their trailing edge. In Fig. 5.28, we study the evolution of the spectral properties of the mid-UV-wavelength peak for pulses of central wavelength increasing from 800 to 825 nm and average IR power 8 mW. The spectrum is peaked at 246 ± 1nm, with maximal rates $S_{246\,max} = 130$ Hz. Although the shape of the peak does vary across the range, its central wavelength seems to be locked at 246 ± 1nm, except for $\lambda_c = 800$ nm (in Fig. 5.28a) and $\lambda_c = 825$ nm (in Fig. 5.28f), for which the mid-UV-wavelength peak is centred at 243 nm ±1nm, and for which the maximal rate is much lower (about 40 Hz). Note that the iris is closed

---

[16]The repeatability of the measurements and scans of the parameter range depends on the mode-locking of the laser and the temporal evolution of the coupling efficiency of the input beams (IR and green) in the fibre. The measurements were barely reproducible at intervals of one hour, and not reproducible from one day to the other.

Fig. 5.27 UV spectrum for a
$P_{\text{avg}} = 8$ mW pulse at
$\lambda_c = 865$ nm interacting
with a 33 mW, 532 nm, CW
probe. Because the iris is
open to a diameter of 15 mm,
the green beam creates
background counts $b_g = 15$
HZ in the UV

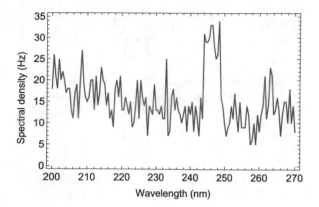

to 8 mm and the noise level (created by the green, IR and laboratory backgrounds)
is $N_{UV} \approx 10$Hz. Thus the SNR for this peak is very good:

$$\text{SNR}_{246} = \frac{S_{246}}{N_{UV}} \approx 13. \tag{5.26}$$

From the spectra in Figs. 5.28 and 5.27, it seems that the peak is relatively insensitive to the central wavelength of the IR pulse. It would be interesting to determine the set of parameters which this mid-UV-wavelength peak depends on such as the probe and pulse powers, relative polarisation of the pulse and probe, interaction length or coupling efficiency of the pulse or probe beams. First, we investigate the dependence on the probe power: the latter is adjusted by controlling the coupling efficiency of the input green beam in the fibre—that is, by lowering this efficiency by means of overlapping more or less the green mode with the cladding (where it is less guided than in the core). In Fig. 5.29, spectra for varying probe power are shown: the peak intensity decreases dramatically with the probe power, and the SNR drops from 13 (for $P_g = 65$ and 60 mW, as in Figs. 5.28 and 5.29, respectively) to 4 for $P_g = 50$ mW (Fig. 5.29b). The peak can barely be seen in Fig. 5.29c for $P_g = 45$, at which point the SNR is just above 1.

Note that no mid-UV-wavelength TH (at 266 nm) was observed on any of the spectra for which the mid-UV-wavelength peak (246 nm) was detected, as exemplified on Fig. 5.29a. We see that, beyond the mid-UV-wavelength peak, there is only one other notable spectral feature: a peak at 303 nm, which corresponds to the long-UV-wavelength TH seen on Fig. 5.26. This indicates that the regime of IR coupling efficiency is such that the 266 nm TH is not phase matched with the IR pulse. The coupling efficiency of the IR might be influenced by the strong green power impinging on the tip of the fibre via thermal effects which would yield a physical deformation of the tip of the fibre such that the IR mode overlaps more or less with the cladding.

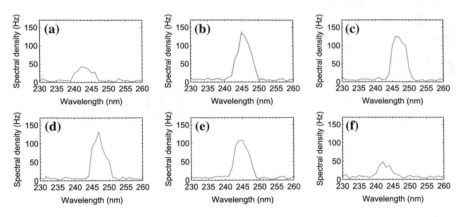

Fig. 5.28 UV spectrum for a 65 mW, 532 nm, CW probe interacting with short-wavelength pulses, for average IR powers of 8 mW. The central wavelength of the pulse is shifted by increments of 5 nm from $\lambda_c = 800$ nm in (**a**) to $\lambda_c = 825$ nm in (**f**). No THG at longer UV wavelengths was observed in any of the 6 configurations. Iris open to 8 mm diameter

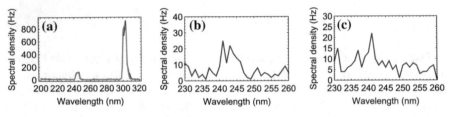

Fig. 5.29 Evolution of the mid-UV-wavelength peak for varying 532 nm probe power and constant, 8 mW, $\lambda_c = 806$ nm, pulse power. **a** Two consecutive measurements for $P_g = 60$ mW are shown in blue and orange (they mostly overlap). The probe power is then lowered to, **b** 50 mW and **c** 45 mW. Note the change in scale of the spectra

Considering the spectra of the 266 nm TH peak for $P_{IR} = 8$ mW on Fig. 5.26, the appearance of the mid-UV-wavelength peak at 246 nm cannot be interpreted as a blueshift of the TH. Indeed, the relative amplitudes are too dissimilar. Furthermore, Eq. (5.25) only allows for red- or blue-shifting of the TH as a function of the fibre dispersion, which is a material property that can most likely not be significantly modified by the CW probe power. Remember that we observed that two TH peaks could be phase-matched with the pulse, and that both of them shifted to longer wavelength, thus a blueshift of more than 20 nm appears extreme, if not impossible. So we cannot explain the generation of the 246 nm peak by means of an obvious nonlinear interaction between the probe and the pulse. Moreover, in [8] the negative-norm signal generated from the soliton was observed around 220 nm, and we predicted similar wavelengths for mode *no*, thus it would be rather surprising if the signal were observed at 246 nm. One could of course wonder whether this peak could be the signal we seek. No conclusive study could be carried to assess this question.

5.4.2.3 Digging in the Unseen

The results presented in the previous section do not allow for the formulation of a firm interpretation of the mid-UV-wavelength peak at 246 nm. Further investigations looking into the effects of the variation of parameters such as the IR power or relative polarisation of the pulse and probe would be necessary to this end. In such investigations, additional optics in the input green beam should be used to allow for its input power to be varied to control the output power, in place of the method used in the present experiments. Indeed, changing the coupling efficiency appears to be a poor method of power management: it is clear that the efficiency of interaction with the IR pulse changes as a function of the overlap of the green mode with the core and the cladding. Likewise, the influence of the coupling of the green beam on the coupling efficiency of the IR beam should be mitigated, for, at present, it clearly has an influence on the higher-order modes in which IR light may propagate, and UV light may be generated.

Now, turning to the sought UV signal, theoretically generated by energy transfer from the positive-norm probe into a negative-norm mode, it is unclear why it has not been detected. Looking back at Eq. (5.23), we found that the various spatial- and frequency-filtering techniques implemented should allow for a signal around 220 nm to be detected with a SNR of at least 1, with a noise of 20 Hz due to the background created (iris diameter: 5 mm). Closing the iris down to 2.5 mm even seems to improve this SNR by a factor 4 at the desired wavelength (see Fig. 5.24d). It might however be that although the SNR for second harmonic generated at 266 nm is increased by closing the slit, it would not be the case for a signal at 220 nm—the short-UV-wavelength component of the beam might diverge strongly at the iris and be filtered out of the beam that reaches the monochromator and single photon counter (SPC), thus reducing the SNR at 220 nm. Indeed it is not a surprise that the SNR is good at 266 nm, for the alignment of the UV beam-path was performed by optimising the rate measured by the SPC at this wavelength (generated by third-harmonic from the IR pulse in the fibre). Actually, one could think of using the 247 nm peak to bring the alignment closer to its optimal settings for short UV-wavelengths. The 220 nm signal in the output of the fibre (with a rate of production of 663 Hz, as in Eq. (5.19)) would ultimately need to be focused down at the iris, so as to allow for closing the latter down to 2.5 mm. Then, according to the calculation (5.20), a rate of 20 Hz should be detected by the SPC. At which point, a $SNR_{220} \approx 4$ would enable for unequivocal detection of the signal.

It might also be that the signal production rate by the interaction of the probe and the pulse in the fibre is lower than in our calculations. In which case, Eq. (5.19) would yield a lower rate. This could be due, for example, to fibre losses (by absorption) in the UV. Let us assume, for the sake of the argument, that an SNR of 1 is the limit of detection. In Sect. 5.4.1.2, we have established that the minimal rate at 220 nm that would allow for detection with a SNR of 1 is ≈ 5 Hz. This is a factor 4 below the rate estimated with no fibre losses. In other words, if less than 25% of the light calculated in Eq. (5.19) would be exiting the fibre, this signal could not be detected. Thus, losses in the UV larger than 5 dB per meter of fibre would suffice to reduce the SNR beyond the limit of detection. If we continue along this line of thought, we arrive at the conclusion that a shorter piece of fibre should be used in order to counter the effects of absorption in the UV. Of course, this would imply reducing the efficiency of the interaction between the probe and the pulse, which linearly depend upon the interaction length $L_{int} = L\nu_{rep}$ by (5.14), with $L = 1.2$ m the fibre length and $\nu_{rep} = 81$ MHz the repetition rate of the laser [1]. Clearly, a compromise between a very short piece of fibre that would not reduce the UV rate by absorption excessively and a minimal interaction length for η_{int} to not reduce η_{no} extremely would have to be found.

Beyond the modification of the fibre length and the filtering of the output, the SNR may be improved by increasing the height of the potential barrier on which the probe scatters. In Sect. 5.4.1.2 we suggested to do so by generating higher order solitons. This could also be achieved by propagating a shorter pulse, down to the few-cycle regime, in the fibre. The change in the refractive index experienced by the probe at the soliton edge would then be extreme even for a fundamental soliton, for few-cycle pulses are extremely steep (with the refractive index varying significantly over a cycle of light). One could also think of using a pulse as the probe to be scattered at the soliton. This circumvents the issue of interaction length, for all the energy in the probe pulse can be made to scatter at the soliton over a very short distance, of the order of a few wavelengths, for adequate group-velocity mismatches.

In conclusion, efforts still have to be put in to dig in the unseen and observe the scattering of a probe of positive-norm to a negative-norm signal at the horizon.

5.5   Conclusion and Discussion

In this section we look back on the experiment presented in this Chapter, and discuss its importance. This discussion basically falls under two main considerations: the intrinsic value of the experiment and its contribution to the field of analogue horizons. We begin with the former by looking out at the route toward the observation of spontaneous emission at an optical horizon, which leads us to comparing the results of the Thesis with investigations of others in the community.

5.5.1   Stimulated and Spontaneous Scattering at the Horizon

The study of the scattering of a probe pulse on a soliton was already suggested in [15]. In this numerical study, the authors interpreted the transfer of energy to the negative-norm mode as a resonant process similar to the generation of NRR, as described in Sect. 2.2.2. However, in the beginning of this section, see Sect. 5.1, we clearly established that the energy transfer is actually due to parametric amplification by means of the scattering of a wave at the horizon and is not a manifestation of ordinary nonlinear fibre optics but a signature effect of analogue horizon physics. In observing this effect, we thus hope to shed some light on the physics of scattering at the horizon. Most importantly, the modes involved in the effect of stimulated scattering also play a role in the effect of spontaneous emission at the horizon: they are *in* and *out* modes of the scattering process, regardless of their incoming state (that is, whether or not they are populated with photons). Therefore, observing the stimulated effect of positive-to-negative-norm scattering will yield essential information about the characteristics of emission from the vacuum.

For example, given the dispersion relation of the fibre in the UV, and the narrow size of the frequency interval over which a front in the refractive index acts as an analogue horizon, the wavelength at which light in the negative-norm mode (the partner in a pair-emission process *à la Hawking*) will be observed should be very similar, if not identical, to that at which the signal in the present experiment would be detected. As in the stimulated case, this negative-norm mode will allow for light to propagate away from the soliton in the left-hand-side region of low refractive index (as in Fig. 5.21). And, in a similar fashion to the stimulated regime, spontaneous emission in the other mode of the pair (light in the mode that allows for it to "escape" the horizon) will be emitted on both sides of the soliton *simultaneously*, but over two distinct wavelength intervals. These two intervals of emission correspond to those over which light from the coherent probe would be red- or blue-shifted, as in Fig. 5.21. Thus these intervals lie on either side of the wavelength at which the probe and pulse would have the same laboratory-frame group-velocity, the velocity-matched wavelength. For the PCF used in the experiment, the central wavelength of the IR pulse that propagates (at $v_g = \frac{2}{3}c$) in the medium at the same group-velocity as the 532 nm probe is $\lambda_c = 835$ nm—see Fig. 5.19: this is the pulse wavelength for which the probe energy does not shift.

5.5.2 *Calculation of Spontaneous Emission Around the Group-Velocity-Matched Wavelength*

In order to articulate the argument of the previous section, it is necessary to know the wavelength and density of emission of light spontaneously emitted from the vacuum. At this point, I refer back to my study of the dispersion relation and emission spectra of bulk fused silica, presented in Chap. 4. For a RIF of height $\delta n = 2 \times 10^{-6}$ (as is created in the experiment) moving at $v_g = \frac{2}{3}c$ in the dispersive medium, the velocity-matched wavelength is $\lambda_m = 396$ nm, and the spectrum of spontaneous emission in the positive-norm modes is as in Fig. 5.30 (the spectrum of emission in the negative-norm mode is shown in Fig. 4.12). The group-velocity-matched wavelength is indicated by a vertical red line and we see that emission into the modes that allow for light to propagate away from the interface is almost symmetrical around λ_m, with short wavelength emission being slightly weaker than long wavelength emission. The discrepancy between the densities at short and long wavelengths are due to the difference in refractive index between the two regions of emission. Indeed, in this

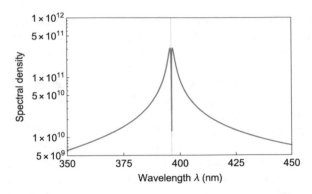

Fig. 5.30 Spectral density of spontaneous emission in the laboratory around the group-velocity-matched wavelength. The spectrum is calculated for a RIF as in Fig. 3.4, of height $\delta n = 2 \times 10^{-6}$, moving at speed $v_{g835} = 1.9992439 \times 10^8$ m.s^{-1} in bulk fused silica. The group-velocity-matched wavelength, $\lambda_m = 396.328 \times 10^{-9}$ nm, is indicated by the vertical red line

calculation, a step geometry for the RIF is considered (as in Fig. 3.4), and thus light at short wavelengths is emitted in mode *uoL* in the high refractive index region. In a real experiment, light at short wavelengths would be emitted in mode *uoL* as well, but this would allow for light to propagate away from the symmetric refractive index profile (as in Fig. 5.21) in the low refractive index region. I expect that the density of emission of light in *uoL* and *moR* would then be strictly identical—this deserves to be thoroughly investigated.

Bermudez and Leonhardt calculated a similar spectrum in [16] in 2016, but did not compare their results with those we had obtained in [17]. They found that the spectrum of emission is indeed symmetrical around λ_m. They studied a simple quadratic dispersion relation, and described the refractive index of the medium by means of a Taylor expansion (as introduced in [1] and developed in [12, 18]). They calculate spectra of spontaneous emission from an extremely short pulse, of $T_0 = 2$ fs, that is a 1.25 cycle-long pulse (for $\lambda_c = 800$ nm) in the laboratory frame. Such a short pulse can probably not be created in an actual experiment, propagate through a dispersive medium or not contaminate the signal frequencies with its own bandwidth. Yet, it is interesting to see that in this limit of pulse-length, the step-like potential provides results close to those obtained with a more complicated refractive index profile.

Resorting to a Taylor expansion of the propagation constant $\beta(\omega)$ to model the refractive index is common in nonlinear optics but cannot provide an accurate perspective on horizon physics. Indeed, our study (based on a Sellmeier model for the refractive index of a dispersive medium) clearly demonstrated that energy transfer between various branches of the dispersion relation may occur (and not only between negative- and positive-optical-frequency branches). This cannot be grasped by the common Taylor-expansion-approach that is restricted to the study of one branch only. Furthermore, the Sellmeier model can be straightforwardly generalised to account for near medium-resonances absorption, which would be helpful in studying the scattering of waves incoming with very high frequencies on the horizon—which is one of the main theoretical unknowns of Hawking's seminal calculation [19] (see also the discussion in Sect. 3.1.2). Thus the present Thesis opens unprecedented opportunities to further investigate (analogue and astrophysical) horizon physics.

5.5.3   The Case for Optical Horizons

The experiment performed in this Thesis is an important step forward in the development of the science of optical horizons. For the optical fibre used in this experiment, the phase-matched wavelength is $\lambda_m = 566\,\text{nm}$, and intervals of spontaneous emission in the positive-norm mode of the pair will lay symmetrically around this wavelength. The emission peaks are expected to be comparatively as narrow and well-defined as those in Fig. 5.30. As was introduced in Sect. 5.5, emission in the UV due to contributions from the negative-norm partner will be concentrated in a very narrow interval similar to the interval over which the signal sought in the experiment should be detected (218–223 nm). Thus, observing the stimulated effect of scattering into this negative-norm mode paves the way to the detection of the spontaneous emission partner, and thus of photon pairs. Indeed, the advantage of the optical setup over other analogue systems (such as water waves [3, 4]) is that it allows for direct, unambiguous, observation of single quanta. Measuring the state of entanglement of the output boils down to measuring correlations in photon-numbers between the negative-norm and positive-norm modes, which are directly accessible quantities, down to the single-photon regime and is not limited to ensemble averages, in optical experiments.

The ability provided by optical experiments to observe single quanta is very important in the case of analogue physics: as we have seen in the previous paragraph, spontaneous emission into the positive-norm mode of the pair will take place over two distinct intervals in the visible simultaneously. In contrast, emission into the negative-norm mode of the pair will be observed over a single, narrow bandwidth, interval in the UV. Thus, quantum correlations between two sets of intervals will have to be measured: the UV-short-wavelength intervals for emission from the analogue white hole horizon, and the UV-long-wavelength intervals for emission from the analogue black hole interval. In both case, knowledge of the interval of emission of the negative-norm partner is key, and the theory and experiment presented in this Thesis is the important step towards identifying it.

In fluid experiments, such as Bose-Einstein Condensates [20–22] or water waves [3, 4], only one of the two analogue horizon configurations can be realised at once[17]: one may only create the analogue to a black- or a white-hole horizon. In optics, we may create both simultaneously, for modes in the vacuum state scatter at both edges of the soliton *continuously*. Furthermore, although the detection of the pair emitted

---

[17]For example, in the waterfall configuration investigated in [21], a black-hole horizon *only* is created. An extra potential barrier would have to be set in the fluid flow to create a white-hole horizon *as well*.

at the horizon is destructive in both fluids and optical experiments (for example, to be detected on a single photon counter, the photon has to be absorbed by the material and cannot be used for anything else subsequently), in the optical case, the quantum state at the output can be used as a resource for other experiments. Indeed, photons are a resource easily handled, even stored, in the laboratory, and can be transferred (or their state teleported [23]) over great distances: provided that one does not perform a destructive measurement (that would be aimed at characterising the quantum state), photons emitted at the horizon can be redirected to other setups. Light spontaneously emitted from the vacuum at the optical horizon thus appears as a new and attractive source of entangled light, for it will be created in a strongly entangled state—very close to a pure state, as we saw in Sects. 3.2.4 and 4.3.2.

References

1. T.G. Philbin, C. Kuklewicz, S. Robertson, S. Hill, F. König, U. Leonhardt, Fiber-Optical Analog of the Event Horizon. Science **319**(5868), 1367–1370 (2008)
2. A. Choudhary, F. König, Efficient frequency shifting of dispersive waves at solitons. Opt. Express **20**(5), 5538 (2012)
3. G. Rousseaux, C. Mathis, P. Maïssa, Thomas G Philbin and Ulf Leonhardt. Observation of negative-frequency waves in a water tank: a classical analogue to the Hawking effect? New J. Phys. **10**(5), 053015 (2008)
4. S. Weinfurtner, E.W. Tedford, M.C.J. Penrice, W.G. Unruh, G.A. Lawrence, Measurement of stimulated hawking emission in an analogue system. Phys. Rev. Lett. **106**(2) (2011)
5. L. Paul Euvé, G. Rousseaux, Génération non-linéaire d'harmoniques après une conversion linéaire en interaction houle-courant. In: *XIVèmes Journées Nationales Génie Côtier—Génie Civil*, ed. by D. Levacher, M. Sanchez et V.Rey (Editions Paralia CFL, Nantes, 2016), pp. 181–190
6. W.G. Unruh, Has hawking radiation been measured? Found. Phys. **44**(5), 532–545 (2014)
7. L.-P. Euvé, F. Michel, R. Parentani, T.G. Philbin, G. Rousseaux, Observation of noise correlated by the hawking effect in a water tank. Phys. Rev. Lett. **117**, 121301 (2016)
8. E. Rubino, J. McLenaghan, S.C. Kehr, F. Belgiorno, D. Townsend, S. Rohr, C.E. Kuklewicz, U. Leonhardt, F. König, D. Faccio, Negative-frequency resonant radiation. Phys. Rev. Lett. **108**(25) (2012)
9. D.E. Spence, P.N. Kean, W. Sibbett, 60-fsec pulse generation from a self-mode-locked ti:sapphire laser. Opt. Lett. **16**(1), 42–44 (1991)
10. J.S. McLenaghan, Negative frequency waves in optics: control and investigation of their generation and evolution. Ph.D. Thesis, University of St Andrews, St Andrews, 2014
11. P. Russell, Photonic crystal fibers. Science **299**(5605), 358–362 (2003)
12. S. Robertson, Hawking radiation in dispersive media. Ph.D. Thesis, University of St Andrews, St Andrews, 2011
13. S. Robertson, U. Leonhardt, Frequency shifting at fiber-optical event horizons: the effect of raman deceleration. Phys. Rev. A **81**, 063835 (2010)
14. G.P. Agrawal, Nonlinear fiber optics. In: *Quantum electronics–principles and applications*, 4th edn. (Elsevier/Academic Press, Amsterdam; Boston, 2007)
15. E. Rubino, A. Lotti, F. Belgiorno, S.L. Cacciatori, A. Couairon, U. Leonhardt, D. Faccio, Soliton-induced relativistic-scattering and amplification. Sci. Rep. 2 (2012)
16. D. Bermudez, U. Leonhardt, Hawking spectrum for a fiber-optical analog of the event horizon. Phys. Rev. A, **93**(5) (2016)

17. M. Jacquet, F. König, Quantum vacuum emission from a refractive-index front. Phys. Rev. A, **92**(2) (2015)
18. S. Robertson, Integral method for the calculation of Hawking radiation in dispersive media. ii. asymmetric asymptotics. Phys. Rev. E **90**(5) (2014)
19. S.W. Hawking, Black hole explosions? Nature **248**(5443), 30–31 (1974)
20. O. Lahav, A. Itah, A. Blumkin, C. Gordon, S. Rinott, A. Zayats, J. Steinhauer, Realization of a sonic black hole analog in a bose-einstein condensate. Phys. Rev. Lett. **105**(24) (2010)
21. J. Steinhauer, Observation of quantum Hawking radiation and its entanglement in an analogue black hole. Nat. Phys. **12**(10), 959–965 (2016)
22. J. Steinhauer, Observation of self-amplifying Hawking radiation in an analogue black-hole laser. Nat. Phys. **10**(11), 864–869 (2014)
23. C.H. Bennett, G. Brassard, C. Crépeau, R. Jozsa, A. Peres, W.K. Wootters, Teleporting an unknown quantum state via dual classical and einstein-podolsky-rosen channels. Phys. Rev. Lett. **70**, 1895–1899 (1993)

Chapter 6
Conclusion

This Thesis consists of the study of the scattering of light at the horizon. Particular emphasis is put on the optical scheme, whereby light in an optical fibre is made to interact with itself to create an effective curvature of spacetime. A theoretical investigation of the motion of light on such a curved background, and of the resulting mixing of waves of positive and negative frequency, is conducted. This mode mixing yields spontaneous emission from the vacuum. An experiment in which an incoming, positive frequency, wave is populated with photons is assembled to observe the transfer of energy from this wave to an outgoing wave of negative frequency at the horizon. This is a classical, stimulated version of the quantum experiment that aims at validating the mechanism of Hawking radiation.

Universality of the Hawking radiation mechanism

Hawking radiation is the late time thermal flux originating from the vicinity of the event horizon of black holes [1, 2]. Any light that propagates through the region of gravitational collapse will experience an exponential gravitational redshift, which implies that the outgoing particles of which black hole radiation is made must be taken to correspond to extremely high frequency radiation at the horizon. Such a Trans-Planckian regime is not described by General Relativity or Quantum Physics—it might be the dominion of Quantum Gravity—and is thus not properly described by the semi-classical approach of Hawking's. This raises questions about the validity of the derivation and the existence of the effect itself.

The fate of Hawking radiation, which is widely seen as a test bench for future theories of Quantum Gravitation, is however not sealed: some laboratory-based systems mimic the kinematics of fields in the vicinity of black holes, and in particular at the event horizon [3]. In total analogy with their astrophysical counterparts, these "dumb holes" will emit a thermal flux. This discovery of Unruh's ushered-in the field of analogue horizons.

Most analogue systems are based on the analogy between the motion of waves in fluids and the motion of waves on a curved background. The problem with Trans-

© Springer International Publishing AG, part of Springer Nature 2018
M. J. Jacquet, *Negative Frequency at the Horizon*, Springer Theses,
https://doi.org/10.1007/978-3-319-91071-0_6

Planckian waves has a direct analogue in fluid systems in terms of the failure of the hydrodynamic limit: it cannot be assumed that perturbations have a wavelength much longer than the healing wavelength and, just like there is no theory for the microphysics of spacetime, there is not one for fluids either. Jacobson suggested modelling the effect of the underlying microphysics on linear fluctuations by considering a modified dispersion relation [4]: he postulated that a modified dispersion relation could be used to understand the breakdown of continuous fluid models due to atomic effects. Unruh then performed numerical simulations in which he demonstrated that, in the presence of dispersion, late-time radiation is not caused by exponentially large quantum fluctuations [5]. This was also applied by Corley to the gravitational case [6]. More recently, Unruh and Schützhold proposed to factor Trans-Planckian effects into the calculation of Hawking radiation via a non-trivial dispersion relation and thus established the universality of the Hawking radiation mechanism [7].

Epistemology of analogue systems

These studies have inspired a number of people who created a large body of theoretical studies of analogue systems, and assembled a handful of experiments to investigate various aspects of analogue horizon physics. Of particular importance is the experiment of Steinhauer who announced having observed correlated emission of phonons at the horizon created in a Bose-Einstein Condensate (BEC) in 2016 [8].

However, further arguments have to be gathered to validate the statement that the observed radiation is of the same class as Hawking radiation. At present, it being the only experiment in which the effect of spontaneous emission has been observed, a BEC with sonic horizon cannot be unequivocally linked to astrophysical black holes. Thus the available evidence is not of the appropriate epistemic type to confirm that black holes do radiate [9]. It lacks external validation.

Other analogue experiments, by means of Unruh and Schützhold universality argument, may rescue the situation: would spontaneous emission be observed in some other setting, confidence in the universality principle would increase and the claim that Hawking radiation is emitted at analogue horizons would be brought closer to validation.

A good candidate scheme is light in an optical fibre. As was mentioned in the beginning of the Conclusion, and demonstrated throughout the dissertation, light in dispersive media can be made to interact with itself and create an analogue event horizon at which spontaneous emission from the vacuum occurs.

Contributions to the field

Spontaneous emission from the vacuum in various systems, and the kinematics and mathematical arguments that support the analogy between laboratory systems and astrophysical black holes are the central problem around which this Thesis has revolved. The contributions of the present work to the endeavours of the community are both theoretical and experimental. They may be deconstructed in the following 5 themes:

- an analytical study of the scattering of light at an interface between two media homogeneous in their refractive index;

- the development of an algorithm to calculate the spectra of spontaneous emission at the interface in all regimes of refractive index change and frequency at the interface, in both the moving and laboratory frame;
- the computation of the first analytical spectra of spontaneous emission as it can be observed in the laboratory frame—including emission in mode solutions of negative frequency;
- an analytical demonstration that emission into a negative frequency mode solution is parametrically amplified when a monochromatic, positive frequency, coherent wave scatters at the interface in the refractive index;
- the experimental investigation of the effect of energy transfer from a monochromatic, positive frequency, continuous wave to positive- and negative frequency waves upon scattering at the horizon created by a soliton in an optical fibre.

The collection of these themes, alongside more general considerations drawn from the state-of-the-art in the field of analogue horizons was presented in this dissertation. For example, the theoretical study[1] yielded the discovery that an interface in the refractive index of an inhomogeneous medium would simultaneously act as a black- *and* white hole emitter, as well as a horizonless emitter, as a function of the frequency and the magnitude of the refractive index change [10]. This is in contrast with the common thought that an interface is either a black- *or* white hole emitter. The implementation of the algorithm allows for direct and fast calculation of spectra, and can be generalised to simulate a variety of refractive index profiles. It thus appears to be a good advance in a field that has been relying on purely numerical studies until recently (see Robertson's work for a similar study to this Thesis [11, 12]). The computation of the spectral density of emission in the laboratory frame, as a result of contributions from all optical modes of positive- *and* negative frequency, yielded the observation that emission was strongest into the negative frequency mode, in the UV. This mode has a negative norm and is the partner in all pair-wise emission process *la Hawking*. Thus it shall be a target of choice in any optical investigation of the spontaneous emission of light from the vacuum. Furthermore, the existence of this UV peak inspired the design of the experiment, that was aimed at detecting energy transferred to this peak from a positive frequency wave upon scattering at the horizon.

Similar experiments have already been carried out in water waves, in which the generation of negative norm waves from the horizon was observed [13, 14]. Moreover, this experiment will yield crucial information toward the observation of correlated photon pairs emitted at the horizon, which is the ultimate signature of spontaneous emission. Observing this emission would contribute to the validation of analogue systems as appropriate source systems to probe the physics of astrophysical black holes.

---

[1]Which included a proof that, in regimes over which the dispersion is linear, the wave equation is analogous to the Painlev-Gullstrand metric—that is, the interface is an analogue horizon.

References

1. S.W. Hawking, Black hole explosions? Nature **248**(5443), 30–31 (1974)
2. S.W. Hawking, Particle creation by black holes. Commun. Math. Phys. **43**(3), 199–220 (1975)
3. W.G. Unruh, Experimental Black-Hole Evaporation? Phys. Rev. Lett. **46**(21), 1351–1353 (1981)
4. T. Jacobson, Black-hole evaporation and ultrashort distances. Phys. Rev. D **44**, 1731–1739 (1991)
5. W.G. Unruh, Sonic analogue of black holes and the effects of high frequencies on black hole evaporation. Phys. Rev. D **51**, 2827–2838 (1995)
6. S. Corley, Computing the spectrum of black hole radiation in the presence of high frequency dispersion: An analytical approach. Phys. Rev. D **57**, 6280–6291 (1998)
7. W. Unruh, R. Schützhold. Universality of the Hawking effect. Phys. Rev. D, **71**(2) (2005)
8. J. Steinhauer, Observation of quantum Hawking radiation and its entanglement in an analogue black hole. Nat. Phys. **12**(10), 959–965 (2016)
9. K.P.Y. Thebault. What can we learn from analogue experiment? arXiv:1610.05028 (2016)
10. M. Jacquet, F. König, Quantum vacuum emission from a refractive-index front. Phys. Rev. A **92**(2) (2015)
11. S. Robertson, Integral method for the calculation of Hawking radiation in dispersive media ii. asymmetric asymptotics. Phys. Rev. E **90**(5) (2014)
12. S. Robertson, U. Leonhardt, Integral method for the calculation of Hawking radiation in dispersive media. i. symmetric asymptotics. Phys. Rev. E **90**(5) (2014)
13. G. Rousseaux, C. Mathis, P. Massa, T.G. Philbin, U. Leonhardt, Observation of negative-frequency waves in a water tank: a classical analogue to the Hawking effect? New J. Phys. **10**(5), 053015 (2008)
14. S. Weinfurtner, E.W. Tedford, M.C.J. Penrice, W.G. Unruh, G.A. Lawrence, Measurement of stimulated hawking emission in an analogue system. Phys. Rev. Lett. **106**(2) (2011)

Appendix A
Positive and Negative Frequency

In this appendix, we will discuss further arguments in favour of the consideration of the sign of frequency of a mode of a field. The discussion will be based on the study of sinusoidal functions.

Sinusoidal functions

A sinusoid is a function of the form

$$x(t) = A \sin(\omega t + \phi), \tag{A.1}$$

where t is an independent (real) variable, and the fixed parameters A (the amplitude), ω (the radian frequency), and ϕ (the initial phase) are all real constants. The argument of the sine function, $\omega t + \phi$ is referred to as the instantaneous phase. Since the sine function is periodic with period 2π, the range of the initial phase is usually restricted to any values between 0 and 2π. The radian frequency ω is the time derivative of the instantaneous phase—$\omega = \frac{d}{dt}(\omega t + \phi)$.

A sinusoid's frequency content may be graphed in the frequency domain by representing its spectral magnitude by (unit-amplitude, and ϕ=0 case)

$$\sin(\omega_x t) = \frac{1}{2i} e^{i\omega_x t} - \frac{1}{2i} e^{-i\omega_x t}. \tag{A.2}$$

That is, the spectrum of a unit-amplitude sinusoid of radian frequency ω_x (and phase $\pi/2$) consists of two components with amplitude 1/2, one at frequency $\omega_x/2\pi$ and the other at frequency $-\omega_x/2\pi$.

Complex sinusoids

We define the complex sinusoid from Euler's Identity

$$e^{i(\omega t + \phi)} = \cos(\omega t + \phi) + i \sin(\omega t + \phi), \tag{A.3}$$

by multiplying it with an amplitude $A > 0$

© Springer International Publishing AG, part of Springer Nature 2018
M. J. Jacquet, *Negative Frequency at the Horizon*, Springer Theses,
https://doi.org/10.1007/978-3-319-91071-0

$$Ae^{i(\omega t+\phi)} = A\cos(\omega t + \phi) + iA\sin(\omega t + \phi). \qquad (A.4)$$

From this equation, we see that a complex sinusoid consists of a real part and an imaginary part—its in-phase and phase-quadrature components, respectively. A complex sinusoid has a constant modulus—a constant complex magnitude.

Given its constant modulus, a complex sinusoid must lie on a circle in the complex plane: a positive-frequency sinusoid ($e^{i\omega t}$, $\omega > 0$) traces out counter-clockwise circular motion along the unit circle as t increases, while a negative-frequency sinusoid ($e^{-i\omega t}$, $\omega > 0$) traces out a clockwise circular motion.[1]

Positive-and negative-frequencies components of a real field

By Euler's Identity, all real sinusoids consist of a sum of opposite circular motion: this is best seen by writing out

$$\sin(\omega t + \phi) = \frac{e^{i(\omega t+\phi)} - e^{-i(\omega t+\phi)}}{2i}. \qquad (A.5)$$

It is obvious that every real sinusoid consists of an equal contribution of positive- and negative-frequency components.[2] Spectrum analysis [1] tells us that every real signal contains equal amounts of positive and negative frequencies. If we denote the spectrum of the real signal $x(t)$ by $X(\omega)$, we have

$$|X(-\omega)| = |X(\omega)|. \qquad (A.6)$$

So why do we usually not consider the negative frequency component of real signals? Well, it is because complex sinusoids have a constant modulus: amplitude envelope detectors (typically, power meters) "compute" the square root of the sum of the squares of the real and imaginary part of the signal to obtain the instantaneous peak amplitude. In other words, we usually convert real sinusoids into complex ones, by removing the negative-frequency component, before processing them.

---

[1] Note that both positive- and negative-frequency sinusoids are necessarily complex.

[2] A real sinusoid is the sum of a positive-frequency and a negative-frequency complex sinusoid.

Appendix B
Modelling of a Change in the Dielectric Constant

In this section of the appendix, we will comment further on the modelling of the dielectric constant in the Sellmeier model, and the modification of this dielectric constant when the refractive index is increased (e.g. by the Kerr effect in the fibre optical experiment of Chap. 5 of this dissertation).

Sellmeier coefficients in the Hopfield model

In Sect. 3.2.3, I explained how the Sellmeier coefficients (the resonance frequency $\omega_i = \frac{2\pi c}{\lambda_i}$ and elastic constant κ_i) of the medium (a set of polarisation fields modelled by harmonic oscillators) were modified by the frequency-dependent change in the refractive index under the step of height δn (as illustrated in Fig. 3.4) by Eq. (3.71). I argued that, in the light of the present lack of theoretical description of the collection of quantum processes from which the dielectric constant of a material arises, the Hopfield model of the dielectric [2] was only an approximation of physical reality. In a scheme based on this approximation, the modulation by (3.71) of the Sellmeier coefficients is a usual description of the change in the dielectric constant within a self-consistent theory.

The question then arises of which of the two coefficients, the resonant frequency or the elastic constant, should be modified to best account for the change in the dielectric constant. I argued that we do not, at present, have at our disposal a good theoretical argument that would discriminate between the two effects—or indeed a combination of both, as proposed in [3, 4] and used in this Thesis and [5].

At the onset of their study, one thus has to make a choice as to which modification to make: either κ_i or ω_i, or both κ_i and ω_i, are modified by the change in the refractive index (of amplitude δn)—note that, in any case, this change has to be frequency dependent. From this choice stem the matching conditions. As I demonstrated in Sect. 3.2.4, if κ_i is to be modified by δn (independently of the modification of ω_i), then there is a discontinuity in the elastic constant at the interface. On the other hand, even if ω_i is modified by δn (independently of the modification of κ_i), this does not create a discontinuity at the interface (the resonant frequency is continuous at the interface for any amplitude of δn). So the choice mentioned above influences the matching conditions, and thus it influences the scattering matrix (because the

© Springer International Publishing AG, part of Springer Nature 2018
M. J. Jacquet, *Negative Frequency at the Horizon*, Springer Theses,
https://doi.org/10.1007/978-3-319-91071-0

elements of the matrix stem from the matching conditions and the amplitude of the modes at the interface).

Dispersion relation

I think that this Thesis demonstrated how the structure and shape of the dispersion relation influence the spectra of emission at the refractive index front (RIF). With this in mind, I would argue that looking at the dispersion relation, which can be readily calculated for any of the three cases we are interested in, may shed some light on the impact of the modification of the Sellmeier coefficients on the spectra of spontaneous emission. For simplicity, we will here focus on the optical branch, and the modes of optical frequency (of positive and negative norm).

Figure B.1 displays the positive-norm optical branch of the dispersion relation in the laboratory frame, and the positive- and negative-norm optical branch in the moving frame (the frame co-moving with the RIF at velocity $u = 0.66c$). The branches are shown in the low- and high-refractive-index region, for $\delta n = 0.048$ at the step. We see that, as only the resonant frequency ω_i is changed by the increase in the refractive index, the distance in the laboratory frame (Fig. B.1a) between the optical branch in the low- and high-refractive-index regions is largest around $k = 0$ and at large $|k|$. In the moving frame (Fig. B.1d), the positive- and negative-norm branches in the high-refractive-index region are very close to their low-refractive-index region counterparts. Thus we expect the moving frame frequency intervals over which the RIF acts as a black-hole and a white-hole to be very narrow. The situation is almost exactly opposite in the case in which only the elastic constant κ_i is changed by the increase in the refractive index (Fig. B.1b), the branch in the high refractive index region overlaps with that in the low refractive index region around $k = 0$ and at large $|k|$, and is furthest away in the medium $|k|$ regime. In the moving frame (Fig. B.1e), the high-refractive-index region branches (of positive- and negative-norm) are far from overlapping with the low-refractive-index region one. Thus we expect the moving frame frequency intervals over which the RIF acts as a black-hole and a white-hole to be relatively large—comparably to the case in which both the Sellmeier coefficients are modified, as discussed in Chaps. 3 and 4 and plotted in Fig. B.1c and f for comparison.

Note that the overall curvature of the dispersion relation does not seem to be affected by the independent change in the Sellmeier coefficients. Thus it is reasonable to assume that the shape of the spontaneous emission spectra will not drastically change either. That is, they should exhibit the same horizon-like features (such as the "shark fin") as those presented in Chap. 4.

Spectra of emission

The algorithm presented in Chap. 4 is based on the matching conditions found in Sect. 3.2.4: it allows for calculating the scattering matrix if the elasticity κ_i is discontinuous, and if the resonant frequency ω_i is continuous, at the interface *only*. This means that, unfortunately, the case of continuous (unchanged) κ_i cannot be investigated without modifying the algorithm, after having found out about the new analytical expression for the matching conditions and relations between Global Modes

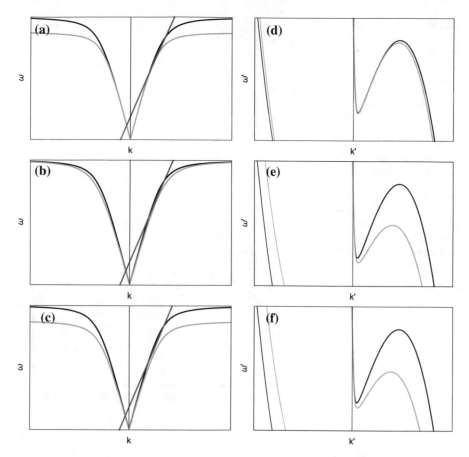

Fig. B.1 Change of elasticity and/or resonance frequency by refractive index increase. Left column: positive-norm optical branch in the laboratory frame; right column: positive- and negative-norm optical branch in the moving frame. The branches are shown in the low refractive index region (black) and high refractive index region (orange), with $\delta n = 0.048$ in bulk fused silica. In (a) and (d), only the resonant frequencies ω_1 and ω_2 are changed; in (b) and (e), only the elastic constants κ_1 and κ_2 are changed; in (c) and (e), both the resonant frequencies ω_1 and ω_2 and elastic constants κ_1 and κ_2 are changed (as in Chaps. 3 and 4). A contour line of constant moving frame frequency ω' is shown on the laboratory frame dispersion plots (a, b and c) to aid the visualisation of the change in the Sellmeier coefficients

(GM) in the inhomogeneous medium. Thus, spectra in the case of unchanged κ_i and changed ω_i cannot be computed presently. On the other hand, it is possible to compute spectra in the case of changed κ_i and unchanged ω_i can be—and the case in which both are affected by the increase in the refractive index was the subject of Sect. 4.3.

I here present new numerical results that allow for comparing the impact of changing both κ_i and ω_i, or changing solely κ_i. The moving-frame photonic flux in the negative-norm optical mode *noL* and the uniquely escaping mode *moR* (that has a

Fig. B.2 Emission spectra
of optical modes of positive-
and negative-norm in the
moving frame in bulk fused
silica. The flux density in
mode *noL* (purple curve) and
moR (orange-dashed curve)
is shown for a step height
$\delta n = 0.048$. In **a** both the
Sellmeier coefficients are
modified (as in Fig. 4.6d) and
in **b** only the elastic constants
κ_1 and κ_2 are modified. The
vertical dotted lines in both
(**a**) and (**b**) indicate the limits
of the subluminal intervals
when both Sellmeier
coefficients are modified by
a $\delta n = 0.048$ (Fig. 4.6d)

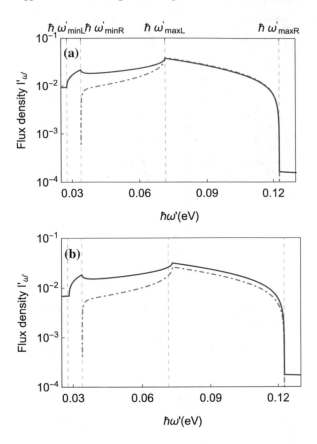

positive norm) are shown in Fig. B.2 when both Sellmeier coefficients are modified
in (a), and when only κ_i is modified in (b). As anticipated in the above paragraph, the
overall shape of the spectra is not significantly affected by the fact that the resonant
frequency is not changed in Fig. B.2b). Horizontal lines indicating the limits of the
subluminal intervals (see Sect. 3.2.3) when both Sellmeier coefficients are modified
are shown for reference. When only the elastic constant is modified, the characteristic
features of the spectra ("shark fin") are slightly shifted to higher frequencies, and it
seems that the emission is overall weaker. Moreover, the emission in *noL* is stronger
than in *moR* over the black-hole interval (between $\hbar\omega'_{maxL}$ and $\hbar\omega'_{maxR}$), which is a
departure from the equality of the fluxes observed when both Sellmeier coefficients
are modified to model the frequency-dependent change in the refractive index.

From these spectra, it is sound to assume that the spectrum of emission when
only the resonant frequency of the medium would be modified would feature the
same shark fin features (and possibly lower flux amplitudes), although over narrower
intervals. The spectrum would then be appreciably different from those presented
on Fig. B.2. Such numerical calculations should definitely be carried to check those
predictions.

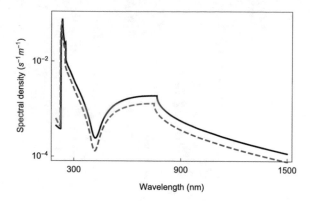

Fig. B.3 Laboratory Spectral Density with modified Sellmeier coefficients. Spectra are calculated for a step of height $\delta n = 0.048$ in bulk fused silica. Dashed-blue: spectrum when κ_1 and κ_2 only are modified; black: spectrum when both Sellmeier coefficients are modified

I also calculated the density of spectral emission in the laboratory frame in the case in which only κ_i is modified to account for the frequency-dependent change in the refractive index under the step of height $\delta n = 0.048$. The LSD is shwon in blue in Fig. B.3: we see that it is mostly similar to the reference spectrum obtained in Sect. 4.3.2, shown here in black. As for the moving frame spectra, the emission is mostly lower when only the elastic constant is modified, and the horizon features are blue-shifted with respect to the reference spectrum.

From these spectra, it is unclear that an optical experiment in glass would allow for clearly distinguishing the best theoretical model (modification of either or both of the Sellmeier coefficients). As I already said in the body of the dissertation, the present theory of the dielectric constant by means of the Hopfield model, and of its modification by an increase in the refractive index by modification of the Sellmeier coefficients has not to be taken too literally: modifying either or both of the elastic constant and resonance frequency is merely a means to account for the frequency dependent change in the refractive index. Unfortunately, the present theory does not account for the collection of quantum processes that would accurately describe the dielectric constant. Further study could be dedicated to extensively studying this puzzle, but for the sake of the present Thesis, modifying both κ_i and ω_i is good enough.

Appendix C
Scattering at a Smooth Profile in a Nondispersive Medium

In this appendix, I demonstrate how to calculate the wave equation for the one dimensional motion of a particle in the field of potential $U(x)$,

$$U(x) = \frac{U_0}{\cosh^2 \alpha x}. \tag{C.7}$$

I then shown how to calculate the coefficients of scattering (reflection and transmission) at such a smooth profile.

Exact solution

The one dimensional motion of a quantum particle in the field of potential $U(x)$ is described by the Schrödinger equation

$$\frac{d^2 \Psi}{dx^2} + \frac{2m}{\hbar^2} (E - U(x)) \Psi = 0. \tag{C.8}$$

In the present case, this reads

$$\frac{d^2 \Psi}{dx^2} + 2m \left(E - \frac{U_0}{\cosh^2 \alpha x} \right) \Psi = 0. \tag{C.9}$$

If we take

$$\xi = \tanh(x), \tag{C.10}$$

where we have implicitly rescaled the problem so that $x \to \alpha x$, then

$$\xi = \frac{\sinh(x)}{\cosh(x)} = \frac{\sqrt{\cosh^2(x) - 1}}{\cosh(x)} \tag{C.11}$$

and

$$\frac{d\xi}{dx} = \xi' = \tanh'(x) = 1 - \xi^2 = \cosh^{-2}(x). \tag{C.12}$$

© Springer International Publishing AG, part of Springer Nature 2018
M. J. Jacquet, *Negative Frequency at the Horizon*, Springer Theses,
https://doi.org/10.1007/978-3-319-91071-0

Thus we get

$$\left(\frac{d}{dx}\right)^2 \Psi = \frac{d}{dx}\frac{d}{d\xi}\Psi\frac{d\xi}{dx} = \frac{d}{dx}\left[(1-\xi^2)\frac{d\Psi}{d\xi}\right] = (1-\xi^2)\left[(1-\xi^2)\frac{d\Psi}{d\xi}\right] \quad (C.13)$$

and Eq. (C.9) can be written as a function of ξ:

$$\alpha^2(1-\xi^2)\frac{d}{d\xi}\left[(1-\xi^2)\frac{d\Psi}{d\xi}\right] + \frac{2m}{\hbar^2}(E - U_0(1-\xi^2))\Psi = 0$$

$$\rightarrow \frac{d}{d\xi}\left[(1-\xi^2)\frac{d\Psi}{d\xi}\right] + \left(\frac{2mE}{\hbar^2\alpha^2}\frac{1}{1-\xi^2} - \frac{2m}{\hbar^2\alpha^2}U_0\right)\Psi = 0. \quad (C.14)$$

Using the replacements $k = \frac{\sqrt{2mE}}{\hbar}$, and $s = \frac{1}{2}\left(\sqrt{1 - \frac{8mU_0}{\alpha^2\hbar^2}} - 1\right)$, and noting that $s(s+1) = -\frac{2mU_0}{\alpha^2\hbar^2}$, the Schrödinger equation (C.14) can be expressed as

$$\frac{d}{d\xi}\left[(1-\xi^2)\frac{d\Psi}{d\xi}\right] + \left(\frac{k^2}{\alpha^2}\frac{1}{1-\xi^2} + s(s+1)\right)\Psi = 0. \quad (C.15)$$

This equation can also be written under a hyper-geometric form by making the substitution $\Psi = (1-\xi^2)^{-\frac{ik}{2\alpha}}\omega(\xi)$. Equation (C.15) becomes

$$\frac{d}{d\xi}\left[(1-\xi^2)\frac{d}{d\xi}\left[(1-\xi^2)^{-\frac{ik}{2\alpha}}\omega(\xi)\right]\right] + \left(\frac{k^2}{\alpha^2}\frac{1}{1-\xi^2} + s(s+1)\right)\left((1-\xi^2)^{-\frac{ik}{2\alpha}}\omega(\xi)\right) = 0. \quad (C.16)$$

The computation of the first term of the above equation yields

$$1^{st}\text{ term} = \frac{d}{d\xi}\left[(1-\xi^2)\left(\xi\frac{ik}{\alpha}(1-\xi^2)^{-\frac{ik}{2\alpha}-1}\omega(\xi) + (1-\xi^2)^{-\frac{ik}{2\alpha}}\omega'(\xi)\right)\right]$$

$$= \frac{d}{d\xi}\left[\xi\frac{ik}{\alpha}(1-\xi^2)^{-\frac{ik}{2\alpha}}\omega(\xi) + (1-\xi^2)^{1-\frac{ik}{2\alpha}}\omega'(\xi)\right]$$

$$= \frac{ik}{\alpha}(1-\xi^2)^{-\frac{ik}{2\alpha}}\omega(\xi) - \xi^2\frac{k^2}{\alpha^2}(1-\xi^2)^{-\frac{ik}{2\alpha}-1}\omega(\xi) +$$

$$+ \xi\frac{ik}{\alpha}(1-\xi^2)^{1-\frac{ik}{2\alpha}}\omega'(\xi) +$$

$$+ (-2\xi)\left(1 - \frac{ik}{\alpha}\right)(1-\xi^2)^{-\frac{ik}{2\alpha}}\omega'(\xi) + (1-\xi^2)^{1-\frac{ik}{2\alpha}}\omega''(\xi). \quad (C.17)$$

Substituting back into Eq. (C.16) and dividing by $\left(1 - \xi^2\right)^{-\frac{ik}{2\alpha}}$ yields

$$-\omega\left(-\frac{ik}{\alpha} - s^2 - s + \frac{k^2}{\alpha^2}\right) - 2\xi\left(1 - \frac{ik}{\alpha}\right)\omega' + \left(1 - \xi^2\right)\omega'' = 0. \qquad \text{(C.18)}$$

Finally, momentarily changing the variable to $u = \frac{1}{2}\left(1 - \xi\right)$, and expressing ω as a function of u - $\omega' = -2\omega'(u)$ and $\omega'' = 4\omega''(u)$ - leads to the Schrödinger equation in its hyper-geometric form

$$u\left(1 - u\right)\omega''(u) + \left(1 - \frac{ik}{\alpha}\right)\left(1 - 2u\right)\omega'(u) - \left(-\frac{ik}{\alpha} - s\right)\left(-\frac{ik}{\alpha} + s + 1\right)\omega(u) = 0. \qquad \text{(C.19)}$$

The exact solution of the problem is the wave function

$$\Psi = \left(1 - \xi^2\right)^{-\frac{ik}{2\alpha}} F\left[-\frac{ik}{\alpha} - s, -\frac{ik}{\alpha} + s + 1, 1 - \frac{ik}{\alpha}, \frac{1}{2}\left(1 - \xi\right)\right] \qquad \text{(C.20)}$$

finite for $\xi = 1$ (i.e. for $x \to \infty$). This solution satisfies the condition that, as $x \to \infty$, $1 - \xi \to 2e^{-2\alpha x}$. Indeed,

$$\left(1 - \xi\right) = \left(1 - \tanh(\alpha x)\right) = \frac{\cosh(\alpha x) - \sinh(\alpha x)}{\cosh(\alpha x)} = \frac{2e^{-2\alpha x}}{e^{\alpha x} + e^{-\alpha x}}, \qquad \text{(C.21)}$$

so for $x \to \infty$,

$$\left(1 - \xi\right) = \lim_{x \to \infty}\left(\frac{2e^{-2\alpha x}}{1 + e^{-\alpha x}}\right) = 2e^{-2\alpha x}. \qquad \text{(C.22)}$$

Physically, this means that as x tends towards infinity, the wave function should only include the transmitted wave (which is proportional to $\exp(ikx)$). The asymptotic form of the wave function as $x \to \infty$ ($\xi \to 1$) (i.e., the wave function before the barrier) is found by transforming the hyper-geometric function [eq:exact solution] with the aid of formula e7 in appendix e of Landau and Lifshitz' book [6].

$$\Psi = \left(1 - \xi^2\right)^{-\frac{ik}{2\alpha}}\left[\frac{\Gamma\left(1 - \frac{ik}{\alpha}\right)\Gamma\left(\frac{ik}{\alpha}\right)}{\Gamma\left(s + 1\right)\Gamma\left(s\right)} F\left(-\frac{ik}{\alpha} - s, s + 1 - \frac{ik}{\alpha}, 1 - \frac{ik}{\alpha}, \frac{1}{2}\left(1 - \xi\right)\right) + \right.$$
$$\left. + \left[\frac{1}{2}\left(1 - \xi\right)\right]^{\frac{ik}{\alpha}} \frac{\Gamma\left(1 - \frac{ik}{\alpha}\right)\Gamma\left(-\frac{ik}{\alpha}\right)}{\Gamma\left(-\frac{ik}{\alpha} - s\right)\Gamma\left(s + 1 - \frac{ik}{\alpha}\right)} F\left(s + 1, -s, 1 + \frac{ik}{\alpha}, \frac{1}{2}\left(1 - \xi\right)\right)\right]$$
$$\text{(C.23)}$$

when $x \to -\infty$, $\xi \to -1$ qnd $F \to 1$. As $\left(1 - \xi^2\right)^{-\frac{ik}{2\alpha}} = \left(1 - \tanh(x\alpha)^2\right)^{-\frac{ik}{2\alpha}} = \cosh(\alpha x)^{\frac{ik}{\alpha}} = \left(\frac{e^{\alpha x} + e^{-\alpha x}}{2}\right)^{\frac{ik}{\alpha}}$, and

$$\left[\frac{1}{2}(1-\xi)\right]^{\frac{ik}{\alpha}} = \frac{(1-\xi)^{-\frac{ik}{2\alpha}}(1+\xi)^{\frac{ik}{2\alpha}}}{2^{\frac{ik}{\alpha}}} = \frac{1}{2^{\frac{ik}{\alpha}}}\left(\frac{1+\xi}{1-\xi}\right)^{\frac{ik}{2\alpha}} = \frac{(e^{\alpha x})^{\frac{ik}{2\alpha}}}{2^{\frac{ik}{\alpha}}}. \quad (C.24)$$

When $x \to -\infty$, $(1-\xi^2)^{-\frac{ik}{2\alpha}} \to \left(\frac{e^{-\alpha x}}{2}\right)^{\frac{ik}{\alpha}} = \frac{e^{-ikx}}{2^{\frac{ik}{\alpha}}}$, and $\left[\frac{1}{2}(1-\xi)\right]^{\frac{ik}{\alpha}} \to \frac{e^{ikx}}{2^{\frac{ik}{\alpha}}}$, which leads to

$$\Psi = \frac{1}{2^{\frac{ik}{\alpha}}}e^{-\frac{ik}{\alpha}}\frac{\Gamma\left(1-\frac{ik}{\alpha}\right)\Gamma\left(\frac{ik}{\alpha}\right)}{\Gamma(s+1)\Gamma(s)} + \frac{1}{2^{\frac{ik}{\alpha}}}e^{\frac{ik}{\alpha}}\frac{\Gamma\left(1-\frac{ik}{\alpha}\right)\Gamma\left(-\frac{ik}{\alpha}\right)}{\Gamma\left(-\frac{ik}{\alpha}-s\right)\Gamma\left(s+1-\frac{ik}{\alpha}\right)}. \quad (C.25)$$

The wave function is composed of the reflected and transmitted waves which are, respectively, the first and second term of Eq. (C.25).

Reflection and transmission coefficients

From Eq. (C.25) it is possible to calculate the reflection coefficient of the potential barrier: it is enough to take the squared modulus of the ration of coefficients in this function -

$$R = \frac{|\Gamma\left(\frac{ik}{\alpha}\right)|^2|\Gamma\left(-\frac{ik}{\alpha}-s\right)|^2|\Gamma\left(s+1-\frac{ik}{\alpha}\right)|^2}{|\Gamma(s+1)|^2|\Gamma(s)|^2|\Gamma\left(-\frac{ik}{\alpha}\right)|^2}$$

$$= \frac{\frac{\pi}{\frac{k}{\alpha}\sin\left(\frac{\pi k}{\alpha}\right)}|\Gamma\left(-\frac{ik}{\alpha}-s\right)|^2|\Gamma\left(s+1-\frac{ik}{\alpha}\right)|^2}{\frac{-s\pi}{\sin(\pi s)}\frac{(-\pi)}{s\sin(\pi s)}\frac{\pi}{\frac{k}{\alpha}\sin\left(\frac{\pi k}{\alpha}\right)}} \quad (C.26)$$

$$= \frac{\sin^2(\pi s)}{\pi^2}|\Gamma\left(-\frac{ik}{\alpha}-s\right)|^2|\Gamma\left(s+1-\frac{ik}{\alpha}\right)|^2.$$

Here can be recognised

$$R = \frac{\sin^2(\pi s)}{\pi^2}\Gamma\left(-\frac{ik}{\alpha}-s\right)\Gamma\left(\frac{ik}{\alpha}-s\right)\Gamma\left(s+1-\frac{ik}{\alpha}\right)\Gamma\left(s+1+\frac{ik}{\alpha}\right)$$

as $\Gamma(z)$ $\Gamma(z')$ $\Gamma(1-z')$ $\Gamma(1-z)$

$$R = \frac{-\sin^2(\pi s)}{\sin\pi\left(\frac{ik}{\alpha}+s\right)\sin\pi\left(\frac{ik}{\alpha}-s\right)}$$

$$R = \frac{-\sin^2(\pi s)}{\left(\sin(\pi s)\cos\left(\pi\frac{ik}{\alpha}\right)+\cos(\pi s)\sin\left(\pi\frac{ik}{\alpha}\right)\right)\left(-\sin(\pi s)\cos\left(\pi\frac{ik}{\alpha}\right)+\cos(\pi s)\sin\left(\pi\frac{ik}{\alpha}\right)\right)}$$

$$R = \frac{-\sin^2(\pi s)}{\cos^2(\pi s)\sin^2\left(\pi\frac{ik}{\alpha}\right)-\sin^2(\pi s)\cos^2\left(\pi\frac{ik}{\alpha}\right)}. \quad (C.27)$$

Finally, the transmission coefficient is $D = 1 - R$

$$D = \frac{\cos^2(\pi s)\sin^2\left(\pi\frac{ik}{\alpha}\right) - \sin^2(\pi s)\cos^2\left(\pi\frac{ik}{\alpha}\right) + \sin^2(\pi s)}{\cos^2(\pi s)\sin^2\left(\pi\frac{ik}{\alpha}\right) - \sin^2(\pi s)\cos^2\left(\pi\frac{ik}{\alpha}\right)}$$

$$= \frac{\cos^2(\pi s)\sin^2\left(\pi\frac{ik}{\alpha}\right) - \sin^2(\pi s)\left[\cos^2\left(\pi\frac{ik}{\alpha}\right) - 1\right]}{\cos^2(\pi s)\sinh^2\left(\pi\frac{k}{\alpha}\right) - \sin^2(\pi s)\cosh^2\left(\pi\frac{k}{\alpha}\right)}$$

$$= \frac{\sin^2\left(\pi\frac{ik}{\alpha}\right)\left[\cos^2(\pi s) + \sin^2(\pi s)\right]}{\sinh^2\left(\pi\frac{k}{\alpha}\right) - \sin^2(\pi s)\sinh^2\left(\pi\frac{k}{\alpha}\right) + \sin^2(\pi s)\cosh^2\left(\pi\frac{k}{\alpha}\right)} \qquad \text{(C.28)}$$

$$= \frac{\sinh^2\left(\pi\frac{k}{\alpha}\right)}{\sinh^2\left(\pi\frac{k}{\alpha}\right) + \sin^2(\pi s)} = \frac{\sinh^2\left(\pi\frac{k}{\alpha}\right)}{\sinh^2\left(\pi\frac{k}{\alpha}\right) + \cos^2\left(\pi\left(s + \frac{1}{2}\right)\right)}$$

$$D = \frac{\sinh^2\left(\pi\frac{k}{\alpha}\right)}{\sinh^2\left(\pi\frac{k}{\alpha}\right) + \cos^2\left(\frac{1}{2}\pi\sqrt{1 - \frac{8mU_0}{\hbar^2\alpha^2}}\right)}$$

f $\frac{8mU_0}{\hbar^2\alpha^2} < 1$, or, in the opposite case,

$$D = \frac{\sinh^2\left(\pi\frac{k}{\alpha}\right)}{\sinh^2\left(\pi\frac{k}{\alpha}\right) + \cos^2\left(\frac{1}{2}\pi\sqrt{\frac{8mU_0}{\hbar^2\alpha^2} - 1}\right)}. \qquad \text{(C.29)}$$

Useful gamma-functions properties

$$\Gamma(s + 1) = s\Gamma(s)$$

$$\Gamma(-s) = \frac{\pi}{\sin(\pi s)\, s\Gamma(s)}$$

$$\Gamma(1 - s)\Gamma(s) = \frac{\pi}{\sin(\pi s)}$$

$$|\Gamma(s)|^2 = \frac{-\pi}{s\sin(\pi s)} \qquad \text{(C.30)}$$

$$|\Gamma(is)|^2 = \frac{\pi}{s\sin(\pi s)}$$

$$\Gamma(s)^* = \Gamma(s^*)$$

if $s \in \mathbb{R}$.

References

1. J.O. Smith, *Mathematics of the Discrete Fourier Transform (DFT)* (W3K Publishing, 2007) http://www.w3k.org/books/
2. J.J. Hopfield, Theory of the contribution of excitons to the complex dielectric constant of crystals. Phys. Rev. **112**(5), 1555–1567 (1958)
3. S. Finazzi, I. Carusotto, Kinematic study of the effect of dispersion in quantum vacuum emission from strong laser pulses. Eur. Phys. J. Plus **127**(7) (2012)
4. S. Finazzi, I. Carusotto, Quantum vacuum emission in a nonlinear optical medium illuminated by a strong laser pulse. Phys. Rev. A **87**(2) (2013)
5. M. Jacquet, F. König, Quantum vacuum emission from a refractive-index front. Phys. Rev. A **92**(2) (2015)
6. L.D. Landau, E.M. Lifšic, Quantum mechanics: non-relativistic theory, in *Course of Theoretical Physics*, vol. 3, ed. by L.D. Landau, E.M. Lifshitz (Elsevier, Singapore, 3). ed., rev. and enl., authorized engl. reprint ed edition, 2007. OCLC: 837367864

Author Biography

Maxime Jules Jacquet

Maxime (born 28 January 1990) is a natural philosopher whose primary research interest is the interplay between Quantum Physics and General Relativity.

In his youth, he enjoyed attending a wide range of public, free state schools until he graduated with a Scientific Baccalauréat in 2008. Still enjoying free tuition, he went on to study engineering at the Polytechnic School of the University of Orléans, France. He spent the academic year 2010–2011 in England via a newly inaugurated partnership between the Universities of Orléans and Staffordshire, UK. Upon returning to France to finish his engineering degree, he decided to specialise in Photonics. In 2013, he went back to the UK to undertake an MSc in Photonics at St Andrews and Heriot-Watt Universities.

That is where he met Friedrich König, with whom he studied toward his Ph.D. at St Andrews between 2013 and 2017. In those three and a half years, he used the tools of Quantum Field Theory on curved spacetime to study the kinematics of waves at optical event horizons in dispersive media. He also conducted experimental investigations of the scattering of classical waves on optical horizons created by few-cycle pulses in photonic crystal fibres. Once more, he was fortunate to be supported in

© Springer International Publishing AG, part of Springer Nature 2018   197
M. J. Jacquet, *Negative Frequency at the Horizon*, Springer Theses,
https://doi.org/10.1007/978-3-319-91071-0

his efforts by public funds, via an EPSRC grant and a 600th Anniversary Fellowship of the University of St Andrews.

From the very onset of his research career, Maxime has considered that the scientific endeavour is twofold: the advancement of knowledge must be shared with others, be it in discussions, teaching and/or outreach activities. He has always enjoyed actively taking part in the teaching activities of his host institution. He invariably found these experience to be enriching, and to be invaluable occasions to learn more about, and reflect on, his teaching as well as his learning philosophy and practices.

After his Ph.D., Maxime's academic peregrinations took him to the University of Vienna, where he opened his research horizons to new maps between theoretical physics and the physical world via foundational studies on time, entanglement, superposition and causal structures. For this young researcher, this is only the beginning of a journey in which experimental and theoretical studies promise to continuously collude.

Publications

2017   Analytical description of spontaneous emission
        of light at the optical event horizon
        with F. König—arXiv:1709.03100
2016   Supercontinuum generation in optical fibers
        with G. Genty, M. Närhi and C. Amiot
        https://doi.org/10.3254/978-1-61499-647-7-233
        Proceedings of the International School of Physics Enrico Fermi
2015   Quantum vacuum emission at a refractive-index front
        with F. König—https://doi.org/10.1103/PhysRevA.92.023851
        Physical Review A **92**, 023851
        Quantum vacuum emission from a moving refractive index front
        with F. König—https://doi.org/10.1117/12.2187987
        Proceedings of SPIE: Quantum Communications and Quantum Imaging
2013   Quantum Vacuum emission at the event horizon
        MSc Thesis, Universities of St Andrews, Heriot-Watt and Orléans

Printed in the United States
By Bookmasters